THE PACIFIC AND ARCTIC OCEANS: NEW OCEANOGRAPHIC RESEARCH

THE PACIFIC AND ARCTIC OCEANS: NEW OCEANOGRAPHIC RESEARCH

KALLEN B. TEWLES
EDITOR

Nova Science Publishers, Inc.
New York

For permission to use material from this book please contact us:
Telephone 631-231-7269; Fax 631-231-8175
Web Site: http://www.novapublishers.com

LIBRARY OF CONGRESS CATALOGING-IN-PUBLICATION DATA

The Pacific and Arctic oceans : new oceanographic research / Kallen B. Tewles, editor.
p. cm.
ISBN 978-1-60692-010-7 (hardcover)
1. Oceanography--Pacific ocean. 2. Oceanography--Arctic ocean. I. Tewles, Kallen B.
GC771.P32 2008
551.46'14--dc22

2008027823

Published by Nova Science Publishers, Inc. ✦ New York

CONTENTS

PREFACE

Global warming and its effects have caused a increased surge of research and funding on ocean research. This new book is dedicted to new oceanographic research on the Pacific and Artic Oceans.

Short Communication A - Hourly measurements of relative humidity, temperature and pressure along with daily precipitation totals recorded at 26 stations in the Canadian Arctic during 1953-2007 were analyzed. The data were first checked for discontinuities. A sudden decrease in relative humidity occurred in the 1970s and 1980s due to an instrument change making it impossible to reliably discern a relative humidity trend due to climate change. A statistically significant warming averaging 5.6 ^{o}C has occurred in winter over the Western Arctic but no change was found in either the Eastern or Northern Arctic. Precipitation varied considerably from year to year and no trends were evident. A statistically significant pressure change averaging -5.8 hPa occurred during winter. The observed temperature increase and pressure decrease during winter may be indicative of changing circulation patterns.

Short Communication B - Paleoclimatic proxies from sedimentary marine sequences often record orbital frequencies (eccentricity, obliquity, and precession) and reveal the effects of insolation on environmental processes. During Pleistocene an important transition occurred in the time interval between 1.25 and 0.7 Ma, the so called Mid-Pleistocene Transition (MPT; Clark et al., 2006) that marked the passage from glacial cycles with 41 to 100-kyr rhythm. In the southwestern Pacific Ocean this transition reflects in paleoceanographic changes as the case east of New Zealand in correspondence of the northward flow of the Deep Western Boundary Current (DWBC). Many studies show evidences of the MPT by the use of proxies dependent on bulk and magnetic sediment grain-size and which provide qualitative information on variations in the strength of the deep Pacific Ocean inflow, possibly directly related to fluctuations of Antarctic Bottom Water production. Many works have been performed about these topics which revealed the importance to improve and enhance the knowledge through future researches.

Chapter 1 - Mexico is one of the 12 megadiverse countries in the world (Mexico, Colombia, Ecuador, Peru, Brazil, Zaire, Madagascar, China, India, Malaysia, Indonesia, Australia). These countries have greater than 70% of the world's biodiversity (McNeely *et al.*, 1990). Mexico has a spectacular 10,000 km of coastline (Farreras, 2006) which is known for its kelp forests, coral reefs, mangroves and coastal lagoons, generally characterized by high productivity, low anthropogenic impact, and high biodiversity. Much of these ecosystems are understudied and it is important to establish baseline data in coastal areas of Mexico as they are threatened by potential (or current) development and pollution which will certainly change the natural fauna and flora of these important areas. The lagoons and coasts of California, USA, are considered endangered habitats due to the loss of 90% of their original area through urbanization (Talley and Ibarra-Obando, 2000). However, neighboring Baja California shows comparatively little anthropogenic impacts to date, yet is faced with the looming threat of rapid and unsustainable development. In this context, an analysis of faunal and environmental conditions of the coastal areas of Mexico is essential for control studies and to establish a baseline.

Chapter 2 - Temporal and spatial variations of vertical profiles of nitrate and phosphate concentrations in seawater in the Kuroshio recirculation (KR) region water columns were analyzed by using a biogeochemical curve fitting method, which includes a function constructed from constant upwelling velocity, and simple biogeochemical processes with the first order kinetics of nutrient consumption and regeneration and particle export flux based on an assumption of steady state. The simple biogeochemical equation was applied to vertical profiles of nutrient concentrations for WOCE P3 (130°-160°E) and Station P9-B (25°N, 137°E) time series data from 1981 to 2002. Most of the vertical profiles of the nitrate and phosphate concentrations can fit well with the biogeochemical equation with correlation factors of >0.998. The spatial distribution of obtained upwelling velocities corresponds to that of meso-scale eddies in the KR region. The analysis of time series data using the curve fitting method provides evidence for inter-annual variations of nutrient profiles as does N:P ratio, which is primarily led by the variability of physical processes such as the variations of generation of meso-scale eddies and its southwest ward motion of meso-scale eddies. The variation of nutrients may be associated with the change of phytoplankton community structure following marine ecological change.

Chapter 3 - While our knowledge about east-west differences in seasonal features of phytoplankton communities in the North Pacific ecosystem has advanced quickly in recent years (i.e., presence of phytoplankton spring bloom in the west while it absent in the east), little information is available on east-west differences in organisms at higher trophic level than phytoplankton. This is partly due to the differences in sampling methodologies for animals at higher trophic levels. In this chapter,, the authors evaluate east-west differences of zooplankton biomass down to the greater depths in the North Pacific employing the same method. They carried out zooplankton sampling with VMPS (60 μm mesh) down to the greater depths (six layers between 0 and 3000 m) at 19 stations between 36˚N and 50˚N along 165˚E and 165˚W in summers of 2003 and 2004. Half of the samples were filtered on 30 μm mesh and used for biomass determination. The other half samples were preserved with 5% borax-buffered formalin and used for microscopic observation. In the land laboratory, biomass determination was done for 8 mass-units (WM, DM, C, H, N, ash, AFDM and Energy). As taxonomic account, large calanoid copepod *Neocalanus cristatus* CV stage were enumerated and their biomass was determined by multiplying separately determined individual mass. Zooplankton biomass integrated over the 0-3000 m depth varied from 5.9 to 28.0 g DM m^{-2}, and was higher at high latitudes. From the viewpoint of east-west comparison, zooplankton biomass in the western (165˚E) stations was 1.7 times higher than that in the eastern (165˚W) stations consistently. As a taxonomic account, biomass of *N. cristatus* CV was greater at western stations (165˚E) than at eastern stations (165˚W), which reflected to the east-west differences in the whole zooplankton biomass. Carbon contents of

N. cristatus CV individuals were higher for the individuals from 165°E stations than those from 165°W stations. These east-west differences in zooplankton biomass (high in west, low in east) may be interpreted by regional differences in phytoplankton abundance of both regions (high in west, low in east).

Chapter 4 - The mechanical behavior (fragmentation and drift) of the sea-ice in the Arctic Ocean is considered in the light of its space and time invariance concluded from the analysis of the geometrical pattern of fragmented sea-ice cover and drift dynamics, respectively. The well-pronounced fractal properties of the Arctic sea-ice cover (ASIC) are interpreted in terms of the concept of self-organized criticality, which implies the perpetual balancing of the open statistical system on the border of stability with multiple cycles of fracturing and restoring occurring at any time and throughout the system. The permanent criticality of this kind needs a dynamic long-range and long-term connectedness of the system, which in the case of ASIC is realized and maintained through a variety of wave and oscillation processes that promote the energy transfer and exchange in the sea-ice cover. An analogy between the energy exchange in the sea-ice drift and the energy release in unstable tectonic formations (earthquakes) is drawn and discussed. The sources of the information were the databases of field observations carried out at the Russian ice-research stations "North Pole", supplemented with the remote technique data available from other origins or reported by other research groups.

Chapter 5 - Variability in the drift of sea ice in the Arctic Ocean is an important parameter that can be used to characterise the thermodynamic processes in the Arctic. Knowledge of the features of sea ice drift in the Arctic Ocean is necessary for climate research, for an improved understanding of polar ecology and as an aid to human activity in the Arctic Ocean. Monthly mean sea ice drift velocities, computed from Advanced Very High Resolution Radiometer (AVHRR), Scanning Multichannel Microwave Radiometer (SMMR), Special Sensor Microwave/Imager (SSM/I), and International Arctic Buoy Programme (IABP) buoy data, are used to investigate the spatial and temporal variability of ice motion in the Arctic Ocean and Nordic seas during the 27 years from 1979 to 2005. Sea ice drift in the Arctic Ocean is characterized by strong seasonal and inter-annual variability. Sea ice drift velocities mirror seasonal changes of the wind in the Arctic, reaching a maximum in December, with a minimum in June. In the central part of the Arctic Ocean and in the area near the Canadian shore the amplitude of this variation is not more than 2 cm s^{-1}. The maximum amplitudes are found in the Fram Strait (9-10 cm s^{-1}), Beaufort Gyre (6-7 cm s^{-1}) and the northern part of Barents Sea (5-6 cm s^{-1}). Low frequency variations of sea ice drift velocities, with periods of 2.0-2.5 yrs and 5.0-6.0 yrs, are related to reorganization of the atmospheric circulation over the Arctic. There is evidence that the average sea ice velocity for the whole of the Arctic Ocean is increasing, with a positive trend for the period 1979-2005. Trends of the monthly mean ice drift velocities are positive almost everywhere in the Arctic Ocean. In the Baffin Bay, Fram Strait and Barents Sea regions, sea ice velocities have increased dramatically, by up to 0.15-0.20 cm s^{-1} per year. The authors suggest that the increase of the sea ice drift velocities in the Arctic Ocean is mostly related to the decrease in both sea ice concentration and ice thickness. The character of the inter-annual variability of ice exchange between the marginal Arctic seas and Arctic Basin and in the Fram Strait in general is similar to the variability of sea ice drift velocities averaged for the whole Arctic Ocean. The authors have found an increase (50-70%) in the ice export from Siberian seas to the Arctic Basin.

In: The Pacific and Arctic Oceans
Editor: Kallen B. Tewles
ISBN: 978-1-60692-010-7

Short Communication A

CLIMATE CHANGE DURING 1953 – 2007 IN THE CANADIAN ARCTIC

W. A. van Wijngaarden[*]
Physics Dept., Petrie Bldg., York University,
4700 Keele St., Toronto, ON Canada, M3J 1P3

ABSTRACT

Hourly measurements of relative humidity, temperature and pressure along with daily precipitation totals recorded at 26 stations in the Canadian Arctic during 1953-2007 were analyzed. The data were first checked for discontinuities. A sudden decrease in relative humidity occurred in the 1970s and 1980s due to an instrument change making it impossible to reliably discern a relative humidity trend due to climate change. A statistically significant warming averaging 5.6 °C has occurred in winter over the Western Arctic but no change was found in either the Eastern or Northern Arctic. Precipitation varied considerably from year to year and no trends were evident. A statistically significant pressure change averaging -5.8 hPa occurred during winter. The observed temperature increase and pressure decrease during winter may be indicative of changing circulation patterns.

I. INTRODUCTION

Evidence showing how human activities are affecting the global climate is steadily accumulating as summarized by the reports of the Intergovernmental Panel on Climate Change [1,2]. This anthropogenic effect is largely due to the rising concentration of greenhouse gases such as CO_2. Climate change is predicted to be more significant in the Arctic than at mid latitudes since rising temperatures will release methane trapped in the permafrost and increase water vapour [3-5]. These two gases are also greenhouse gases and therefore further amplify the warming. This in turn may increase relative humidity and

[*] E-mail: wlaser@yorku.ca

precipitation as well as affect weather circulation patterns. Hence, climate observations in the Arctic may very well act as the "early warning canary" of climate change. Indeed, the extent of Arctic sea ice during 2007 was the smallest ever recorded in modern history [6-8].

The Canadian Arctic archipelago juts into the Arctic Ocean. The station of Alert on Ellesmere Island, located at 82.5° N latitude, is significantly closer to the North Pole than northern Alaska and Siberia. Hourly and daily observations of climate have been made since the 1950s at a number of stations in the Canadian Arctic and are available in digital form from Environment Canada. These measurements are taken at the surface and are more extensive than the observational record of the upper atmosphere. The dataset also covers a longer period than is available from satellite observations. The latter must be calibrated using ground based station measurements. This can be problematic as shown recently in the Antarctic where temperatures as determined from satellite observations were found to be in error by as much as 10 °C [9]. These station observations are also an essential input to global circulation models that attempt to interpolate the climate between observing stations [10]. This is obviously a challenge in the Canadian Arctic which encompasses an area of several million square kilometers where "neighbouring" stations are separated by several hundred or even a thousand kilometers.

This paper is organized as follows. First, the extent of the data record and the limitations of the various measurements are discussed. It is essential that effects due to changing instruments and/or procedure be understood to allow the proper determination of climate trends. An example is given showing the effect on relative humidity caused by the replacement of the psychrometer by the dewcel. Next, observations of temperature, precipitation and pressure are discussed and trends are plotted. Finally, conclusions are presented.

II. Data Analysis

Hourly observations of relative humidity, temperature and pressure along with daily observations of precipitation were retrieved from the Environment Canada archive for the 26 stations listed in Table 1. Most of these stations were opened in the 1950s. Archival records show each station received regular inspections to ensure that instruments were properly calibrated and that data taking procedures were properly implemented. Table 1 shows the percentage of missing hourly observations was less than 10% for most stations. Only three stations, Alert, Cape Hooper and Longstaff Bluff have slightly greater than 50% missing data. Most of the missing hourly data occurred in the 1950s and early 1960s when data were only measured every 6 hours at some stations.

Figure 1 shows the total number of measurements recorded each year. This number increases as additional stations were opened in the 1950s and 1960s. There was a slight decrease in the mid 1990s due to budget cuts that affected station operations. The station density in the Arctic is lower than in Southern Canada. Nevertheless, the total number of measurements analyzed in this study exceeds 10 million, which is considerable.

Data was analyzed as follows. Average values were computed for the seasons defined as follows: winter (December – February), spring (March – May), summer (June – August) and fall (September – November). For the case of hourly data, averages were also found for the

Table 1. Station List. The percentage of missing data is rounded to the nearest 5%. A star denotes daily precipitation records that were available and analyzed in this study.

Province/ Territory	Name	Station Number	Latitude	Longitude	Hourly Data	
					Start Date	% Missing
Yukon	Burwash	2100182	61.37	139.17	Oct. 1, 1966	15
	Mayo	2100700	63.62	135.87	Jan. 1, 1953	15
	Watson Lake*	2101200	60.12	128.82	Jan. 1, 1953	<1
	Whitehorse*	2101300	60.72	135.07	Jan. 1, 1953	<1
NWT	Cape Parry	2200675	70.17	124.72	Jan. 1, 1956	5
	Fort Simpson	2202101	61.77	121.23	Jan. 1, 1953	<1
	Fort Smith*	2202200	60.02	111.97	Jan. 1, 1953	<1
	Hay River*	2202400	60.83	115.78	Jan. 1, 1953	<1
	Inuvik*	2202570	68.3	133.48	Jan. 1, 1953	<5
	Norman Wells*	2202800	65.28	126.8	Jan. 1, 1953	<5
	Yellowknife*	2204100	62.4	114.43	Jan. 1, 1953	<1
Nunavut	Baker Lake*	2300500	64.3	96.08	Jan. 1, 1953	13
	Coral Harbour*	2301000	64.2	83.37	Jan. 1, 1953	<5
	Alert*	2400300	82.52	62.28	Jan. 1, 1953	57
	Cambridge Bay*	2400600	69.1	105.13	Jan. 1, 1953	<5
	Cape Hooper	2400660	68.28	66.48	Mar. 1, 1956	60
	Clyde	2400800	70.48	68.52	Jan. 1, 1953	15
	Eureka*	2401200	79.98	85.93	Jan. 1, 1953	35
	Hall Beach*	2402350	68.78	81.25	Jan. 1, 1953	<5
	Iqaluit*	2402590	63.75	68.55	Jan. 1, 1953	<1
	Longstaff Bluff	2402684	68.54	75.09	Dec. 1, 1955	55
	Resolute*	2403500	74.72	94.98	Jan. 1, 1953	10
Manitoba	Churchill*	5060600	53.97	101.1	Jan. 1, 1953	<1
Quebec	Inukjuak	7103282	58.47	78.08	Jan. 1, 1953	30
	Kuujjuarapik*	7103536	55.28	77.77	Jan. 1, 1953	<1
	Kuujjuaq*	7113534	58.1	68.42	Jan. 1, 1953	<5

night (0 – 5 am), morning (6 – 11 am), afternoon (noon - 5 pm) and evening (6 – 11 pm). No discernible difference between trends observed during these four different periods of the day were found for relative humidity, temperature and pressure. A linear function was then fit to each time series given by

$$y_t = a + bt + e_t \tag{1}$$

where y_t is the seasonal value observed at the station of interest at time t and e_t is the residual. A t-test then determined whether the slope or trend given by b was statistically significant at the 5% confidence level.

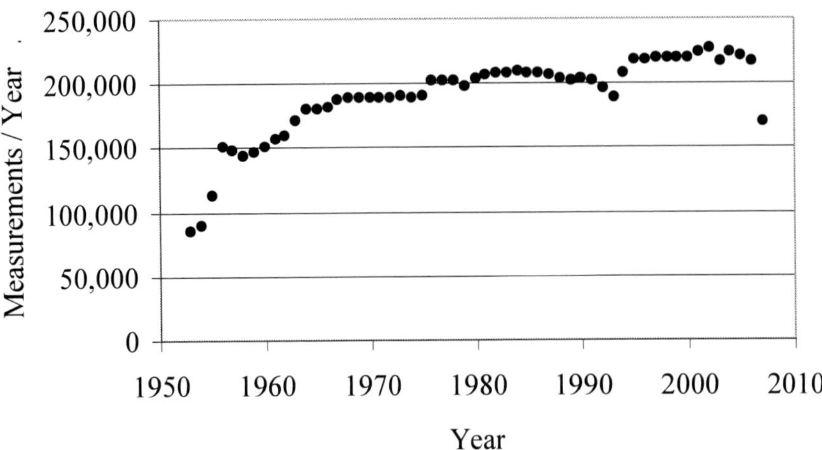

Figure 1. Number of Hourly Measurements per year taken at 26 stations in this study. The number of measurements increased in the 1950s and 60s as stations were opened. The small dip in the early 1990s reflects budget cuts which resulted in staff reductions. The number is anomalously low in 2007 because only data taken before Oct. 2007 were considered.

a) Relative Humidity

Examination of the relative humidity data found a discontinuity for nearly every station. Figure 2 shows the relative humidity decreased suddenly in 1970 at Hay River, with the largest decrease of about 20% occurring in winter. Data was examined for step inhomogeneities such as shown in Figure 2 by fitting the time series by

$$y_t = a + bt + cI_t + e_t \tag{2}$$

where I_t equals 1 for $t \geq p$ and zero otherwise. The value of p providing the minimum residual sum of squares then determines the most probable year of a potential step [11]. The F statistic was then used to determine whether the data is fit better using either equation (1) or (2) [12].

Nearly all the Arctic stations were found to have step discontinuities occurring sometime during the 1970s or 1980s [13]. The station histories showed the date of the observed discontinuity coincided with the replacement of the pscyhrometer by the dewcel. For Hay River, this instrument change occurred in 1970. The psychrometer measures the temperature difference between a wet and a dry bulb thermometer. At very low temperatures, the wet bulb freezes rapidly and the temperature difference if any, is very small. This leads to anomalously high values of relative humidity as are evident in Figure 2.

Data observed at a given station can be corrected for inhomogeneities by comparing time series obtained at neighbouring stations. This has worked well for correcting temperature time series for stations that are in close proximity in southern Canada [14]. In the case of the Arctic, stations are separated by large distances and considerable caution should be exercised when attempting this procedure [15].

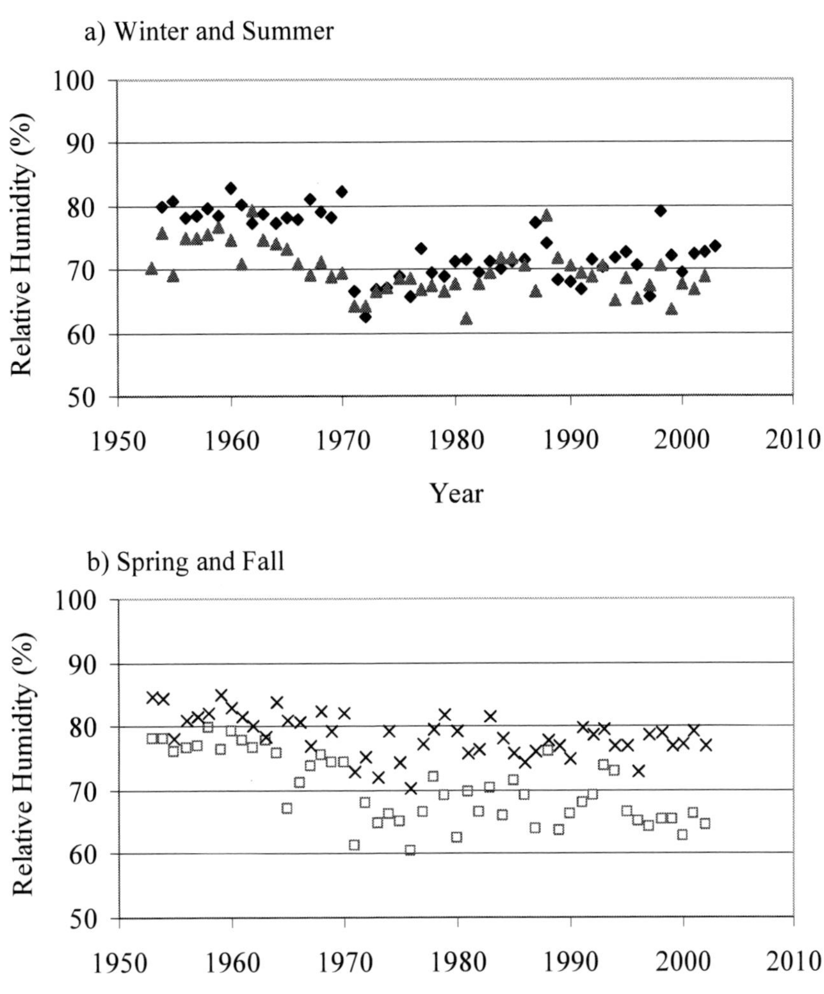

Figure 2. Relative Humidity at Hay River, North West Territories during a) winter (solid black diamond) and summer (filled red triangle) as well as b) spring (open red square) and fall (black cross). The apparent reduction in relative humidity in 1970 coincided with the replacement of the psychrometer by a dewcel as discussed in the text.

b) Temperature

Temperature was measured with a resolution of 0.1 °F prior to 1977 [16]. These measurements were later recorded in the archive with 1 °F precision. Beginning in 1977, the Celsius temperature scale was adopted and temperature was observed with a resolution of 0.1 °C. The earlier archival measurements were then converted to Celsius and recorded with a precision of 0.1 °C. Examination of daily [14] and hourly [17] measured temperatures have found relatively few inhomogeneities. Moreover, these discontinuities have a much smaller magnitude than was the case for the relative humidity data.

Figure 3 shows the average temperatures recorded in the various seasons at four stations located in the southern (Yellowknife), western (Inuvik), eastern (Iqaluit) and northern Arctic

(Alert). Trend lines were fit to the seasonal time series and are shown in Figure 4. Few statistically significant trends are in evidence anywhere during spring, summer and fall. However, in winter all 11 stations located in the Yukon and Northwest Territories show statistically significant warming trends averaging +5.6 °C over the period 1954-2007. The three stations further east, Cambridge Bay, Baker Lake and Churchill also report statistically significant trends in winter averaging +2.7 °C. This is consistent with previous work that found a warming of about 5 °C during winter on the Canadian prairies during 1954-2003 [18, 19]. The result for the Western Arctic in winter differs from that found for stations in the northern and eastern Arctic where no statistically significant trends occur. This winter warming is consistent with the observation of markedly less ice observed in the Arctic Ocean north of Siberia and Alaska in late summer [6-8]. It appears that the winter warming produces less thick ice which in turn melts faster during the summer.

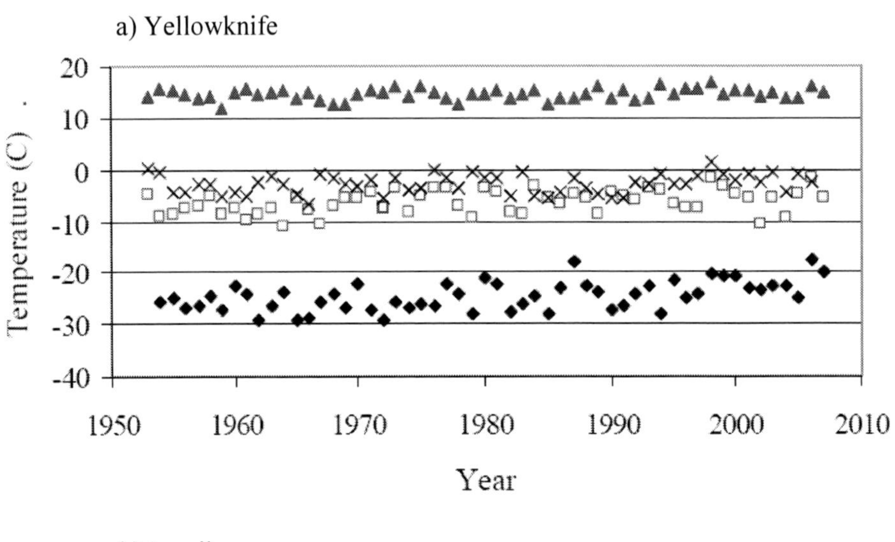

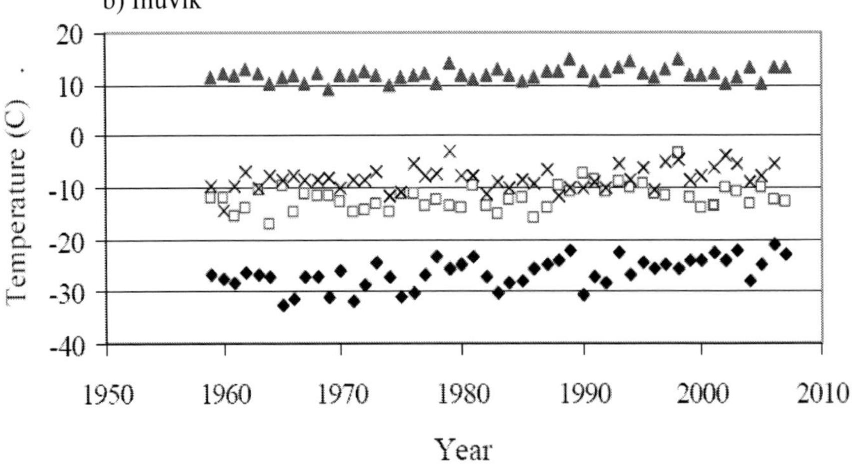

Figure 3. (Continues)

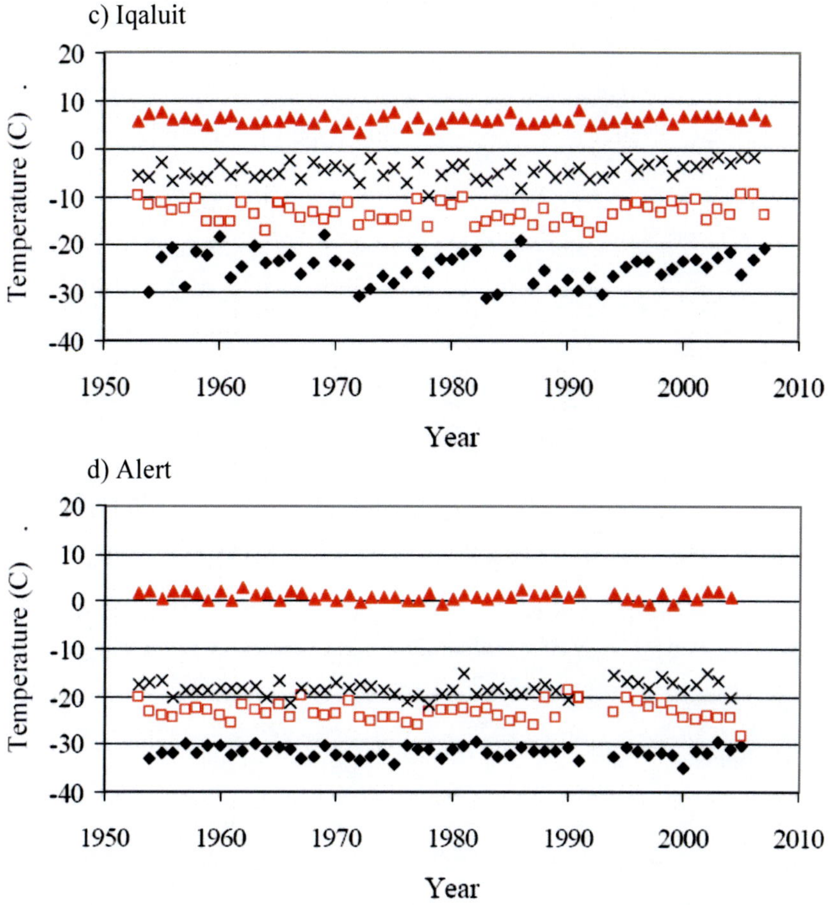

Figure 3. Seasonal Dependence of Temperature at a) Yellowknife, b) Inuvik, c) Iqaluit and d) Alert for winter (solid black diamond), spring (open red square), summer (solid red triangle) and fall (black cross). The gap in the Alert data in the mid 1990s resulted from cutbacks in station operation due to budget reductions.

c) Precipitation

Daily precipitation is recorded in units of 0.1 mm in the climate archive. A number of factors affect the accuracy of these data. First, missing data can have a larger effect when estimating the total precipitation compared to determining the average temperature. The amount of missing daily data is believed to be less than 5% at most stations. This is somewhat difficult to estimate as sometimes zero precipitation may be recorded on a day when no measurement was made. A second uncertainty results from the introduction of new gauges that has affected the collection efficiency of precipitation by 5% or more in southern Canada [20]. Precipitation measurement in the Arctic is further complicated because a significant fraction of the annual total is received in the form of trace amounts of snow or rain (< 0.3 mm) whose cumulative total is hard to reliably estimate. Distinguishing between blowing snow and new snow fall is also a challenge. Finally, the snow moisture content differs from station to station making it difficult to estimate the so called snow water equivalent [20].

These uncertainties are significant over much of the Arctic which receives less than 25 cm annual precipitation and is therefore a desert.

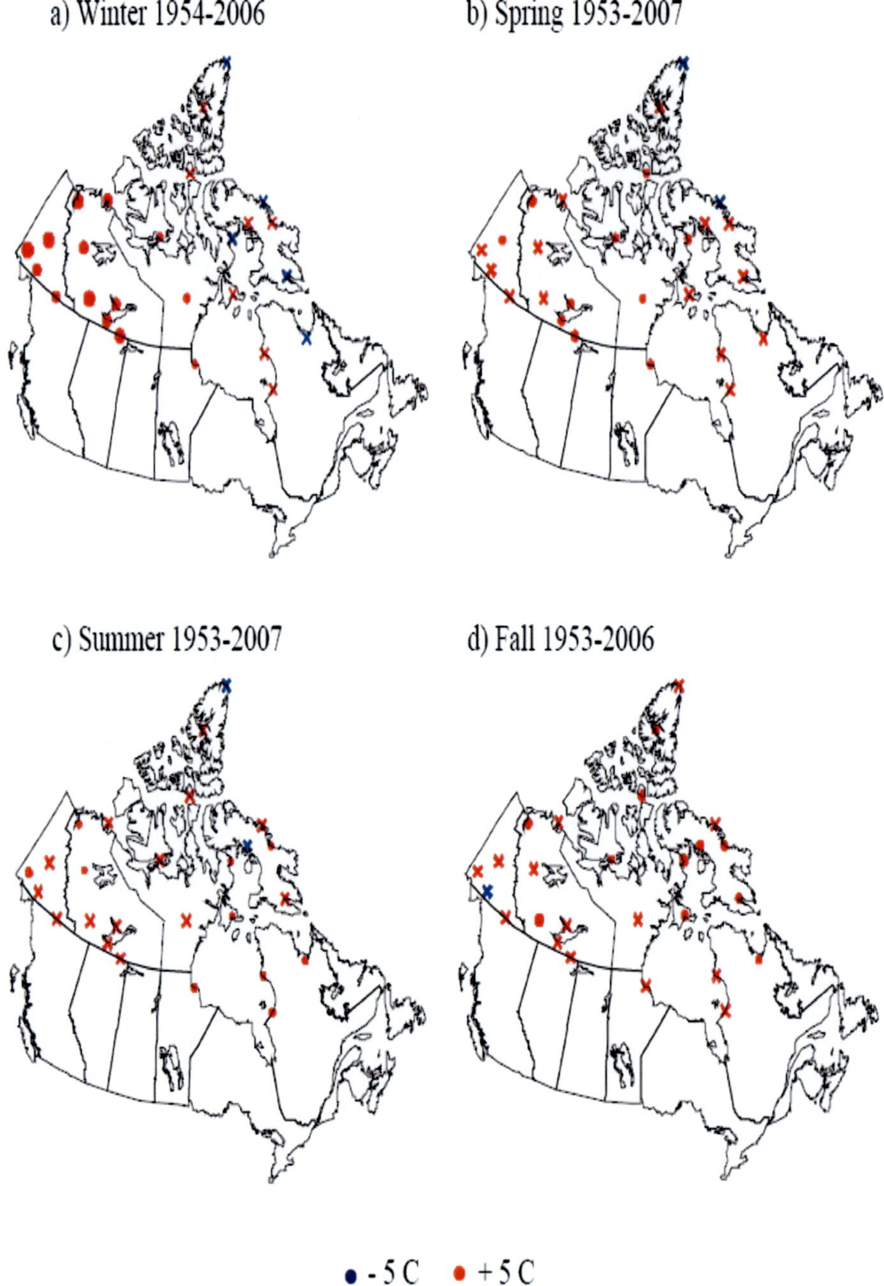

Figure 4. Temperature trends during a) winter, b) spring, c) summer and d) fall. Red (blue) dots represent increasing (decreasing) temperature trends statistically significant at the 5% level. Crosses represent insignificant trends.

Figures 5 and 6 show the precipitation at Alert and Yellowknife in the different seasons. Most of the precipitation is received during summer and fall. This is not surprising as the air contains little moisture at very cold winter temperatures. Similarly, southern stations such as Yellowknife that experience warmer temperatures than northern Arctic stations such as Alert, receive more precipitation. Figures 5 and 6 show considerable year to year variation in precipitation. Hence, it was not possible to discern any trend at a given station. The average precipitation experienced by the stations indicated in Table 1 was therefore computed to check for any regional trend.

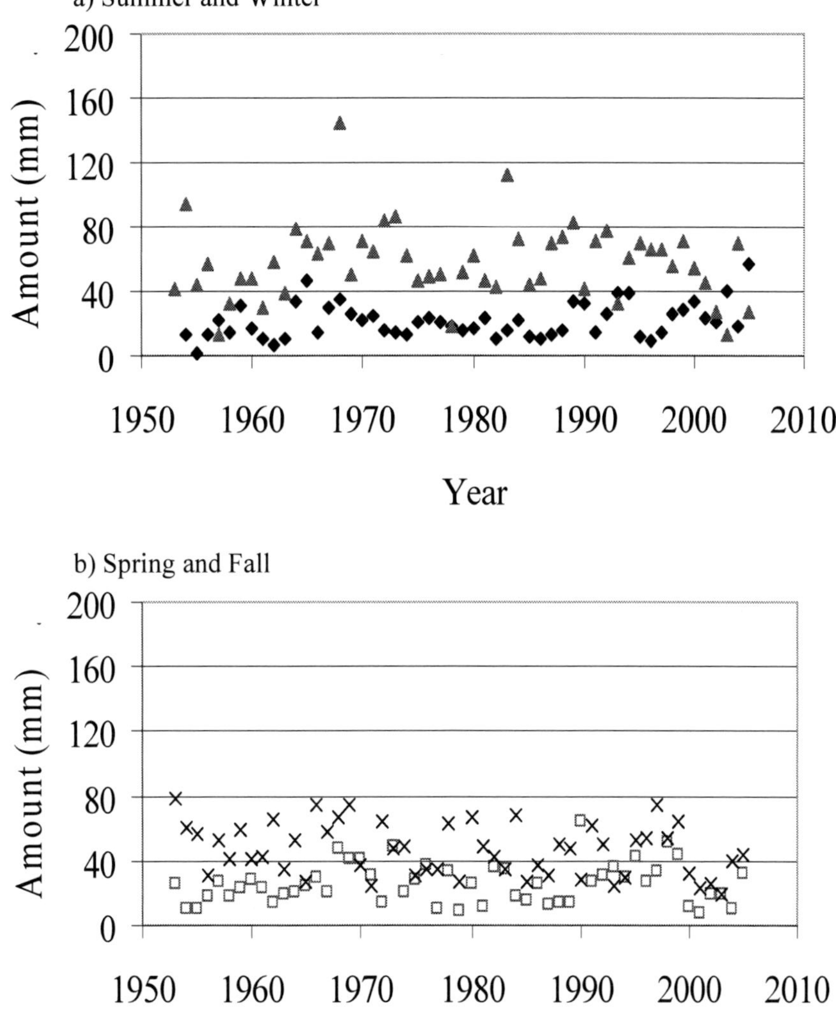

Figure 5. Precipitation at Alert during a) Summer (solid red triangle) and Winter (solid black diamond) and b) Spring (open red square) and Fall (black cross).

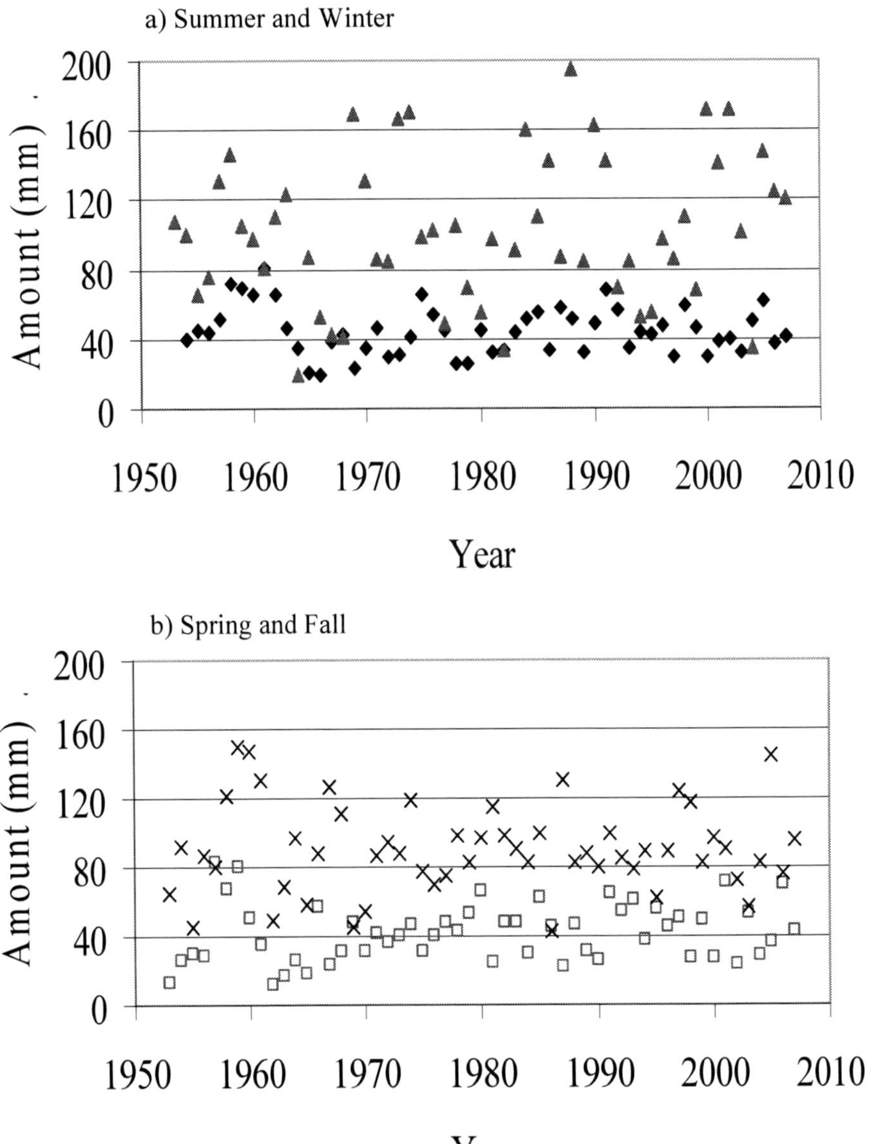

Figure 6. Precipitation at Yellowknife during a) Summer (solid red triangle) and Winter (solid black diamond) and b) Spring (open red square) and Fall (black cross).

Figure 7 shows the average precipitation per station observed for all the stations and separately for stations located in the northern Arctic i.e. Nunavut. The northern Arctic clearly receives less precipitation in every season. Figure 7 does not show evidence of either increasing or decreasing precipitation in any season. These results are consistent with studies for southern Canada that have not found any significant change in precipitation except on the prairies in the winter where a reduction of about 50% has been observed [19, 21].

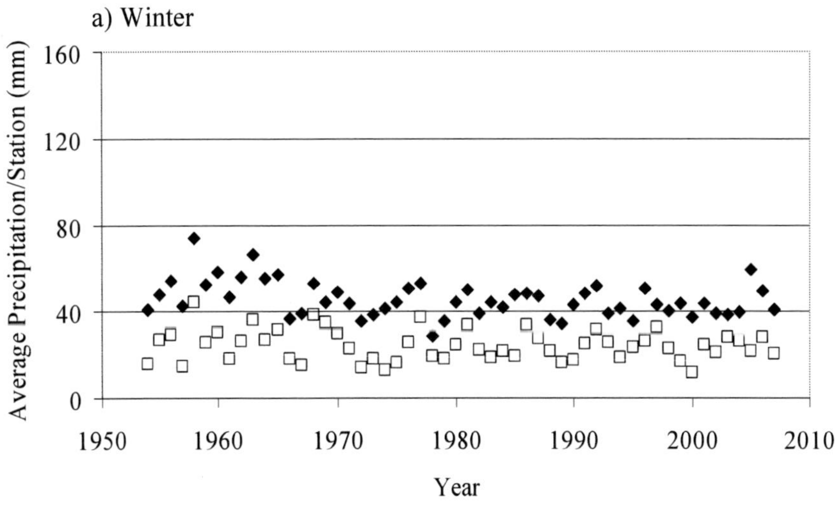

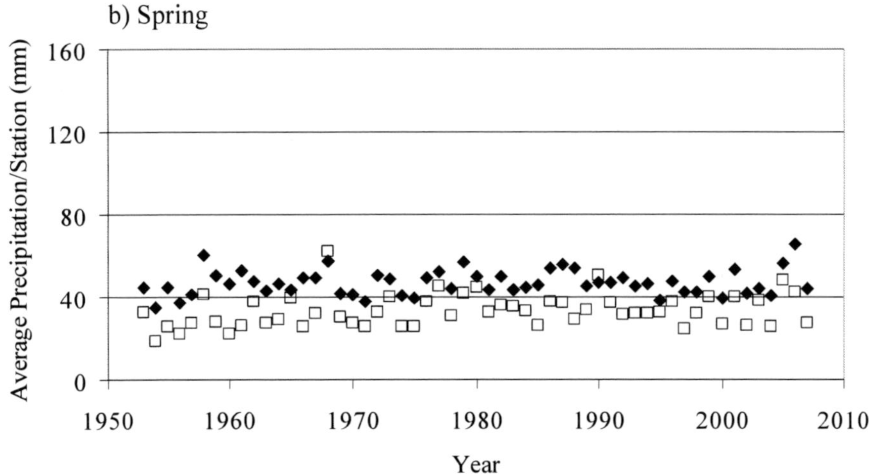

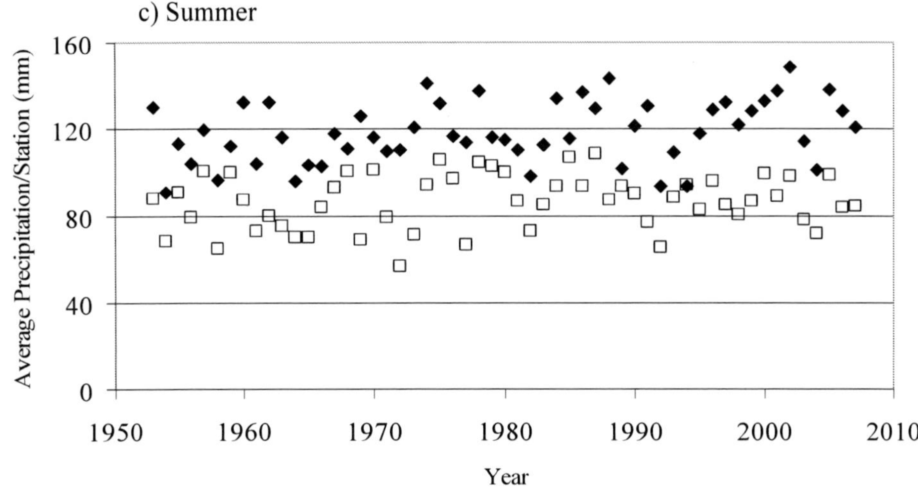

Figure 7. (Continues)

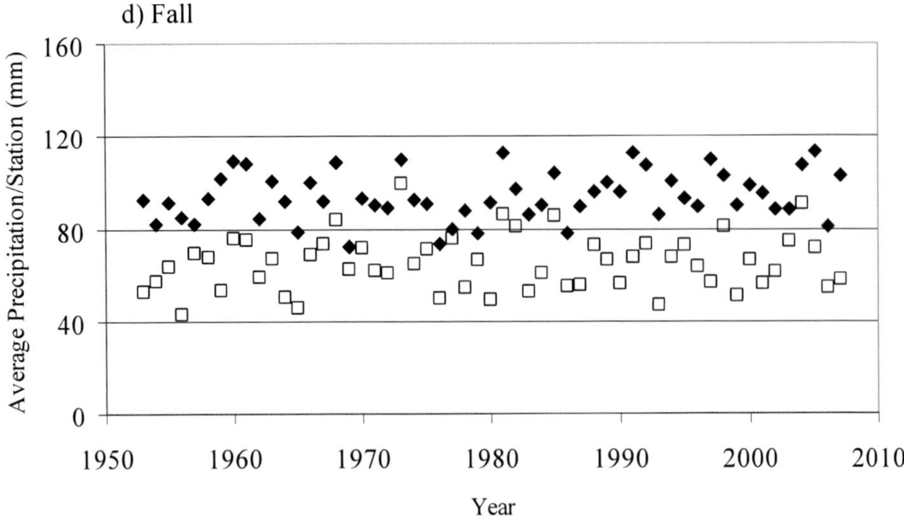

Figure 7. Average Annual Precipitation per Station during a) winter, b) spring, c) summer and d) fall. Solid black diamonds were found for all stations as indicated in Table I while the empty squares represent only data measured at stations located in Nunavut.

d) Pressure

Pressure was recorded with an accuracy of 0.1 hPa using a mercury manometer [22]. The pressure obviously depends on the station altitude which has been taken into account by some studies that consider the adjusted sea level pressure. Unfortunately, a variety of errors have been found in the records of adjusted sea level pressure that arise from changing estimates of the station altitude [23, 24]. In contrast, the station pressure measurements have been found to be relatively free of inhomogeneities [23].

Figure 8 shows the pressure observed at Inuvik in the different seasons. There is scatter from season to season but a statistically significant downward trend occurs in winter. Figure 9 shows the pressure trends observed over the entire Arctic. Eighteen of the 26 stations report a

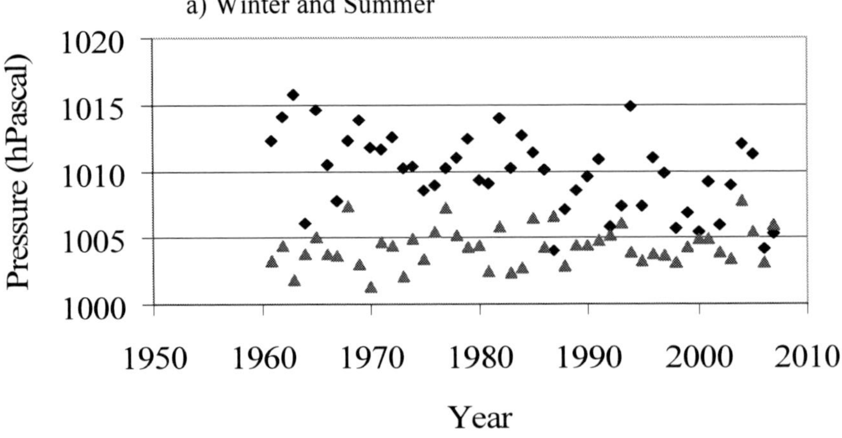

Figure 8. (Continues)

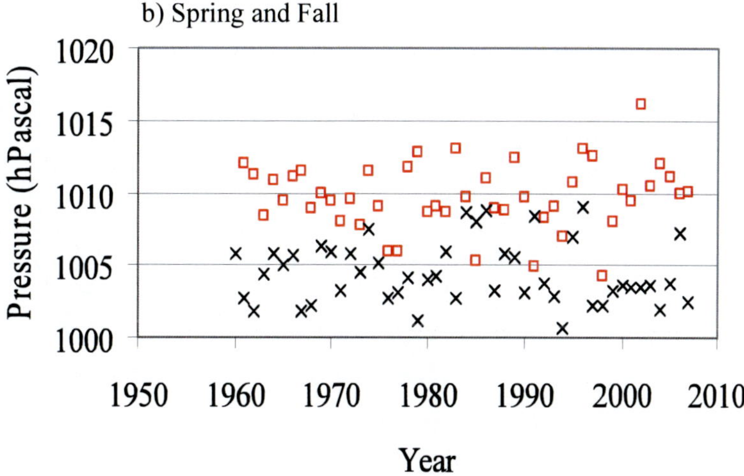

Figure 8. Seasonal Dependence of Pressure at Inuvik for a) winter (solid black diamond) and summer (solid red triangle) and b) spring (open red square) and fall (black cross).

statistically significant pressure decrease in winter averaging 5.8 hPa over the 1954 to 2007 period. The number of stations reporting decreasing pressure trends in the other seasons is fewer and the magnitude of these pressure changes is smaller. These results are comparable to a pressure decrease of as much as 4 hPa that has been found by another study over parts of the Arctic during winter in the 1968-1997 period [25, 26].

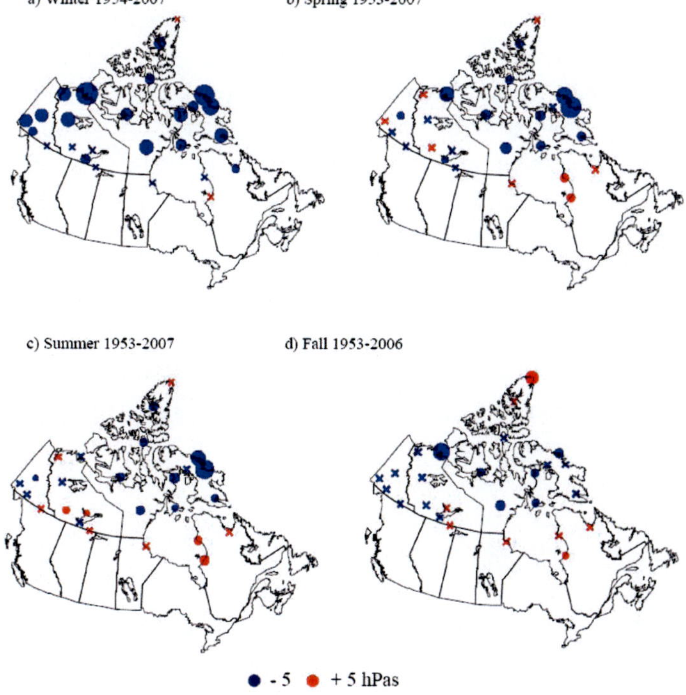

Figure 9. Pressure trends during a) winter, b) spring, c) summer and d) fall. Red (blue) dots represent increasing (decreasing) pressure statistically significant at the 5% level. Crosses represent insignificant trends.

The pressure reduction in winter may be a result of a strengthening of the so called North Atlantic Oscillation (NAO) whose strength is determined by the difference in pressure over Iceland relative to the Azores [27]. A strong NAO indicates a strong high pressure center over the Azores and anomalously low pressure over Iceland. Trends of a stronger NAO in recent decades are believed to have reduced the severity of winter weather over most middle and high latitude Northern Hemisphere continental regions [28, 29]. The NAO may be part of a much larger oscillation called the Arctic Oscillation (AO) that affects the entire Northern Hemisphere [30, 31]. The AO has recently been linked to Arctic Sea Ice variation [32].

III. CONCLUSIONS

This study shows warming is occurring in the Canadian Arctic. However, the data reveal the warming is primarily affecting the Western Arctic and only in the winter season. The average observed temperature increase of 5.6 ^{o}C in the Yukon and Northwest Territories over the past half century is very rapid when compared to geologic time scales. Hence, an anthropogenic cause for this heating is likely. The close relation between this temperature increase and the average decrease of pressure of 5.8 hPa found at 18 of the 28 stations is intriguing. This could be a clue that the winter warming has resulted from changing weather circulation patterns. Indeed, there has been speculation of an anthropogenic effect on pressure [33].

It is important that these observed changes in the Arctic climate be understood to comprehend how the global climate will be altered in the coming century. It would be useful to analyze archival measurements of wind speed and direction. These do exist but caution must be exercised as the calibration of these instruments must be checked frequently to ensure their performance does not degrade in the harsh Arctic environment. Moreover, surface wind measurements are very sensitive to the local topography and easily perturbed by a change in surrounding structures. Upper atmosphere measurements of wind would be more useful but are not readily available for as extensive a time period as hourly surface measurements.

It will be especially interesting to see if statistically significant changes of temperature become evident in the future in the seasons other than winter and whether the eastern and northern Arctic also begin to experience winter warming. These changes could also be accompanied by increased precipitation since warmer air can hold more moisture. Air moisture may be expected to further increase due to increased evaporation from the unfrozen Arctic Ocean. Hence, continued monitoring of the Arctic climate is essential to improve our understanding of global climate change in the 21st century.

ACKNOWLEDGMENTS

The author would like to thank the Natural Science and Engineering Research Council of Canada for financial support and Environment Canada for providing access to the archival climate data.

REFERENCES

[1] Intergovernmental Panel on Climate Change (2001), Climate Change 2001: The Scientific Basis: Contributions of Working Group to the Third Assessment Report of the Intergovernmental Panel on Climate Change, edited by J. T. Houghton et al, 881 pp., Cambridge Univ. Press, Cambridge, U.K. and IPCC Report (2001).

[2] Climate Change 2007 – The Physical Science Basis (Working Group I), Impacts, Adaptation and Vulnerability (Working Group II), and Mitigation of Climate Change (Working Group III), Intergovernmental Panel on Climate Change Cambridge Univ. Press, Cambridge, UK (2007).

[3] R. Macdonald, "Awakenings in the Arctic", *Nature*, 380, 286-287 (1996).

[4] M. Sturm, D. K. Perovich, and M. C. Serreze, "Meltdown in the North", *Scientific American* 289, 60-67 (2003).

[5] L. Bengtsson et al, "The Early Twentieth Century Warming in the Arctic – A Possible Mechanism", *J. Climate* 17, 4045-4057 (2004).

[6] M. C. Serreze et al, "A Record Minimum Arctic Sea Ice Extent and Area in 2002", *Geophys. Res. Lett.* 30, No. 3, 1110 (2003).

[7] M. Johannessen et al, "Arctic Climate Change: Observed and Modelled Temperature and Sea Ice Variability", *Tellus,* 56A(4), 328-341 (2004).

[8] S. Renfrow, National Snow and Ice Data Center, www.arctic.noaa.gov (2007).

[9] J. R. Docken, "Antarctic surface temperature comparison study on satellite approximations and land-based measurements", *Proceedings of American Meteorological Society Meeting,* New Orleans, (2008).

[10] E. Kalnay et al "The NCEP/NCAR 40-year Reanalysis Project", *Bull. Am. Meteor. Soc.* 77, 437-471 (1996).

[11] L. A. Vincent, "A technique for the identification of inhomogeneities in Canadian temperature series", *J. Climate* 11, 1094-1104 (1998).

[12] X. L. Wang, Comments on "Detection of undocumented changepoints: A revision of the two-phase regression model", *J. Climate*, 16, 3383-3385 (2003).

[13] W. A. van Wijngaarden and L. A. Vincent, "Examination of Trends in Hourly Surface Relative Humidity in Canada during 1953-2003", *J. Geophys. Res.* 110, D22102 (2005).

[14] L. A. Vincent, X. Zhang, B. R. Bonsal and W. D. Hogg, "Homogenization of daily temperatures over Canada", J. Climate 15, 1322-1334 (2002).

[15] L. A. Vincent, W. A. van Wijngaarden and R. Hopkinson, "Surface Temperature and Humidity Trends in Canada for 1953-2005", *J. Climate* 20, 5100-5113 (2007).

[16] Environment Canada Observation Manual, "*Guide to Meteorological Instruments and Methods of Observation*", 6th Edition World Meteorological Organization ISBM 9263160082 (1976).

[17] W. A. van Wijngaarden, "Investigation of Hourly Records of Temperature, Dew Point, Relative Humidity and Specific Humidity for the Analysis of Climate Trends in Canada", *Environment Canada Contract KM040-4-5118 Report* (2005).

[18] X. Zhang, L. A. Vincent, W. D. Hogg and A. Niitsoo, "Temperature and Precipitation Trends in Canada during the 20th Century", *Atmos. Ocean* 38, 395-429 (2000).

[19] W. A. van Wijngaarden and L. A. Vincent, "Trends in Relative Humidity in Canada from 1953-2003", *Proceedings of American Meteorological Society Meeting*, Seattle (2004).

[20] E. Mekis and W. D. Hogg, "Rehabilitation and analysis of Canadian daily precipitation time series", *Atmos. Ocean* 37, 53-85 (1999).

[21] L. A. Vincent and E. Mekis, "Changes in daily and extreme temperature and precipitation indices for Canada over the twentieth century", *Atmos. Ocean* 44, 177-193 (2006).

[22] Atmospheric Environment Service Canada. 1961, 1970, 1976. "*Manual of Standard Procedures and practices for Weather Observing and Reporting*".

[23] W. A. van Wijngaarden, "Examination of Trends in Hourly Surface Pressure in Canada during 1953-2003", *Int. J. Climatol.* 25, 2041-2049 (2005).

[24] V. C. Slonosky and E. Graham, "Canadian pressure observations and circulation variability: links to air temperature", *Int. J. Climatology* 25, 1473-1492 (2005).

[25] J. E. Walsh, W. L. Chapman and T. L. Shy, "Recent decrease of sea level pressure in the Central Arctic", *J. Climate* 9, 480-486 (1996).

[26] L. C. Nkemdirim and D. Budikova, "Trends in sea level pressure across western Canada" *J. Geophys. Res.* 106, 11801-11812 (2001).

[27] J. W. Hurrell, "*The North Atlantic Oscillation: Climatic Significance and Environmental Impact*", American Geophys. Union, Washington D. C. USA 279 (2003).

[28] D. W. J. Thompson and J. M. Wallace, "Regional Climate Impacts of the Northern Hemisphere Annular Mode", *Science*, 293, 85-89 (2001).

[29] G. M. Ostermeier and J. M. Wallace, "Trends in the North Atlantic Oscillation – northern hemisphere annular mode during the twentieth century", *J. Climate* 16, 336-341 (2003).

[30] D. W. J. Thompson and J. M. Wallace, "The Arctic Oscillation Signature in the Wintertime Geopotential height and Temperature Fields", *Geophys. Res. Lett.* 25, 1297-1300 (1998).

[31] M. H. P. Ambaum, B. J. Hoskins and D. B. Stephenson, "Arctic oscillation or North Atlantic oscillation?", *J. Climate* 14, 3495-3507 (2001).

[32] G. Rigor, J. M. Wallace and R. L. Colony, "Response of sea ice to the Arctic oscillation", *J. Climate* 15, 2648-2663 (2002).

[33] N. P. Gillett, F. W. Zwiers, A. J. Weaver and P. A. Stott, "Detection of human influence on sea-level pressure", *Nature* 422, 292-294 (2003).

In: The Pacific and Arctic Oceans
Editor: Kallen B. Tewles
ISBN: 978-1-60692-010-7

Short Communication B

RECORDS OF CLIMATE AND PALEOCEANOGRAPHIC VARIABILITY DURING THE MID-PLEISTOCENE TRANSITION IN THE PACIFIC OCEAN

A. Venuti[1], M. Cobianchi[2] and C. Lupi[2]
[1]Istituto Nazionale di Geofisica e Vulcanologia, Roma, Italy
[2]Dipartimento di Scienze della Terra, Università degli Studi di Pavia, Pavia, Italy

ABSTRACT

Paleoclimatic proxies from sedimentary marine sequences often record orbital frequencies (eccentricity, obliquity, and precession) and reveal the effects of insolation on environmental processes. During Pleistocene an important transition occurred in the time interval between 1.25 and 0.7 Ma, the so called Mid-Pleistocene Transition (MPT; Clark et al., 2006) that marked the passage from glacial cycles with 41 to 100-kyr rhythm. In the southwestern Pacific Ocean this transition reflects in paleoceanographic changes as the case east of New Zealand in correspondence of the northward flow of the Deep Western Boundary Current (DWBC). Many studies show evidences of the MPT by the use of proxies dependent on bulk and magnetic sediment grain-size and which provide qualitative information on variations in the strength of the deep Pacific Ocean inflow, possibly directly related to fluctuations of Antarctic Bottom Water production. Many works have been performed about these topics which revealed the importance to improve and enhance the knowledge through future researches.

1. THE MID-PLEISTOCENE TRANSITION (MPT)

The knowledge of the Earth's climate system, in terms of components and evolution, is in continuous progress and bring to the realization that human activities are now beginning to alter the climate and, in particular, that they are increasing the speed of processes that were already acting in the past but with a different timing. Help in explaining these topics, above all the complex interactions between biosphere and climate system, comes from the study of a peculiar time span of change and arrangement of the climate system and oceanographic circulation that leads to the present conditions. Particularly, the time interval, spanning from 1.25 to 0.7 Ma (Clark et al., 2006) and known as "Middle Pleistocene Transition (MPT)", represents a critical interval for studying paleoceanographic evolution coupled with timing and cause-effect relationships between biosphere and climate system. It is characterized by considerable global changes both in the climate system and in the oceans, accompanied by a general decrease in greenhouse gases (Raymo 1997; Tziperman and Gildor, 2003). During the MPT glacial-interglacial periodicity switches from 41ky to 100ky (Berger and Jansen, 1994) and a major "super-interglacial", characterized by strong oscillations in the amplitude of insolation cycles (Teitler et al., 2007), occurs at about 1 Ma (Marine Isotope Stage 31). The most relevant modifications developing during the MPT include ice-sheet expansion, strong sea-level falls (Prell, 1982; Kitamura and Kawagoe, 2006), increased aridity and intensity of African and Asian monsoons (Clemens et al., 1996; Sun et al., 2006), increased erosion on continental shelf, and consequent enhanced organic carbon flux into the deep-sea sediments. In addition, modifications in the ocean circulation and specifically in the intensity of thermohaline circulation ("conveyor belt", Boyle and Keiwing 1985; Venz and Hodell 2002), in nutrient and alkalinity seawater distribution (Boyle 1988) are well documented features at the MPT.

Evidences in the sedimentary record of Pleistocene climate variations may be inferred using various proxies as $\delta^{18}O$ and $\delta^{13}C$ and other geochemical parameters (Sr/Ca, Mg/Ca, Cd/Ca, Zn/Ca, Cd/P) (Boyle, 1981; Shackleton et al., 1990; Lea et al., 2000; Marchitto et al., 2002). These proxies allowed to reconstruct the sea surface and deep water variations as response to intensity of the glacial-interglacial cycles and related oceanographic events (e.g. changes in the global thermohaline circulation system). The biosphere was also influenced by the environmental changes occurred during the MPT: major extinction events (10 species per 0.1 Ma) are recorded in the benthic foraminiferal deep-sea records between 0.7 and 0.58 Ma ("*Stilostomella* Extinction", Hayward, 2001; 2002; Kawagata et al., 2005; 2006) as a response to climate deterioration. Planktonic foraminiferal assemblages show significant turnover through MIS 23-22 (Zheng et al., 2005) and some of the taxa, such as the modern form of *Neogloboquadrina pachyderma* (sinistral) appeared in this time interval (about 1 Ma, Kucera and Kennett, 2002). Calcareous nannofossil assemblages modified in conjunction with sea surface temperature and circulation. An increase in their diversity is recorded between 1.2 and 0.6 Ma, as a response to a weaker activity of the North Atlantic Deep Water (NADW), and consequently to changes in the nutrient availability from 0.9 My (e.g., Bassinot et al. 1997). Abundance fluctuations of some calcareous nannofossil genera (e.g., *Gephyrocapsa, Helicosphaera, Calcidiscus, Syracosphaera*), appear to be related to ocean circulation linked to glacial-interglacial cycles (Flores and Sierro, 2007).

2. Evidences of MPT in Eastern New Zealand Oceanic Sedimentary System

The study of sedimentary sequences in the area east of New Zealand revealed insight of the response of the Earth's climate system during the MPT. Eastern New Zealand Oceanic Sedimentary System (ENZOSS) model represents this area (e.g., Carter et al., 1996a; 2004), which extended from the Solander Channel north to the Kermadec Trench and where different channels influx from the continental margin, from the south, Solander, Bounty, and Hikurangi Channel that at present is the major sediments contributor to the Deep Western Boundary Current (DWBC, Carter et al, 2004). Part of the sediment after deposition on a "fan-drift" is subducting into the Kermadec Trench in the north (Carter and McCave, 1994; Carter et al., 2004). Sedimentary material is transported by the main currents, the DWBC from the south and the East Cape Current from the north-west. Chatham Rise represents an important morphological feature in this area, it extends west to east for c. 1300 km and is in correspondence of the Subtropical Convergence (Fenner et al., 1992; 2004; Nelson et al., 1993; 2000; Nodder et al., 2003). On the sides of the Rise there are terraces and off the northern side sedimentary drifts formed at deeper depths, where there are actually slow currents <10 cm s-1 (Carter et al., 1996b; McCave and Carter, 1997; Carter and McCave, 1997). High resolution paleoclimatic records comes from the study of the Site MD97-2114 and ODP Site 1123 which are under the influence of the upper and middle part of the DWBC, respectively (Hall et al., 2001; Venuti et al., 2007). The DWBC has actually a significant average volume transport (16±11.9 106 m^3s^{-1} at 32°30'S) (Whitworth et al., 1999); the passage in this area represents the gateway to the Pacific Ocean, for this reason it is likely that changes in paleoceanography are recorded in sedimentary sequences from these locations.

We refer for example the data from MD97-2114 Site located at 42°22'27"S, 171°20'42"W on the northern flank of the Chatham Rise above the carbonate compensation depth. This marine sedimentary sequence consists in homogeneous mud with abundant calcareous nannoplankton and foraminifera and several tephra layers related to the Quaternary volcanism in New Zealand. A paleomagnetic parameter, the ARM/IRM ratio, sensitive to variations in magnetite grain-size in the sediment, has been used as paleocurrent proxy; its long-term trend indicates higher bottom current flow during the last 300 ka and between ca. 1.06 - 0.87 Ma (Venuti et al., 2007).

Among the microfossil organisms, the calcareous nannofossils observed in the site MD 97-2114 and ODP 1123 record the well known evolutive first and last appearance events (Raffi, 2002; Raffi et al., 2006), together with other eco-biostratigraphic bioevents (de Kaenel et al., 1999; Flores et al., 1999; 2000; Baumann and Freitag, 2004; Maiorano and Marino, 2004) such as the abundant acme events of the genus *Gephyrocapsa*. Moreover in the site MD972114 close-spaced events of calcareous nannofossils have been recorded (from MIS 29 to MIS 25) falling within the MPT (Lupi 2007, PhD Thesis). These paleontological data have been correlated with the $\delta^{13}C$ curve that shows two major shifts in the MIS 28-25 and in the MIS12-11 intervals (Lupi 2007, PhD Thesis) and seem to precede the two major climate change and ice-sheet expansion events in the Pleistocene, i.e. the Middle Pleistocene Transition, and the mid-Brunhes event, in agreement with Wang et al. (2003). These evidences open new perspectives to investigate in detail the relationship between biological evolution and climate changes.

Physical, isotope and biotic paleoclimatic proxies from sites at different water depths (MD97-2114 at 1935 m and ODP Site 1123 at 3290 m) show evidences for an external forcing mechanism that drives the strength and water mass properties of the DWBC during the MPT. Although the sediment supply in this basin depends on many factors and at least three sources from the south and west should be considered (Venuti et al., 2007), the physical proxies (ARM/IRM and the sortable silt mean size, respectively) clearly record changes in the strength of the DWBC flow. However both these studies (Hall et al., 2001 and Venuti et al.,

2007) pointed out the weak character of the DWBC or better of the thermohaline circulation during the MPT.

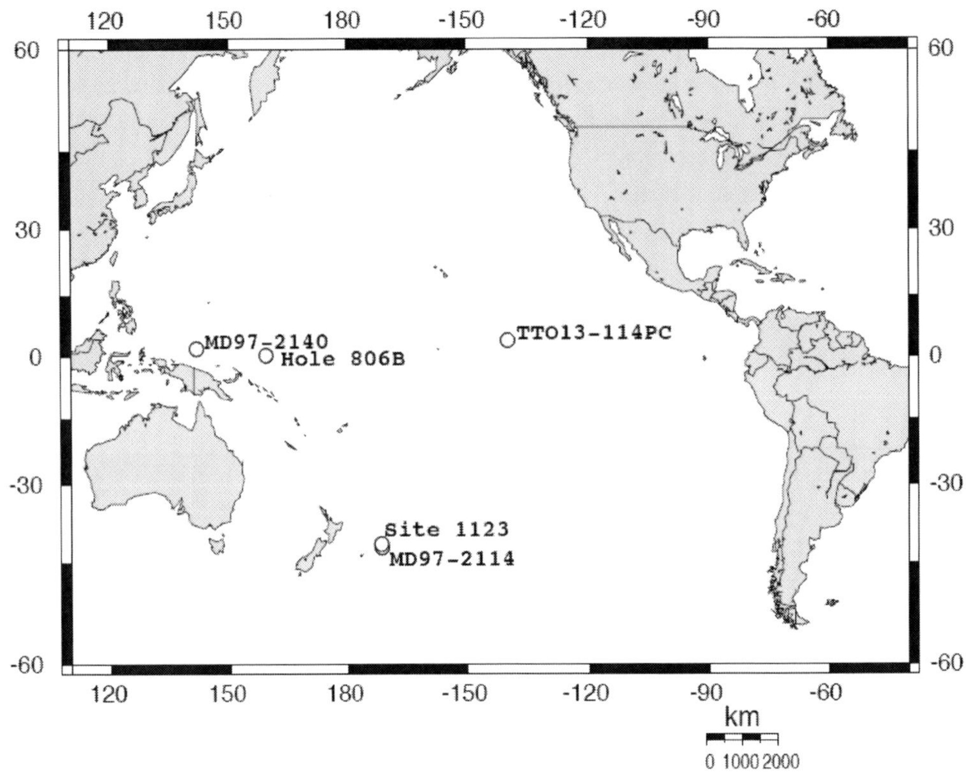

Figure 1. Locations of the Sites discussed in the text.

3. Other Insights from Marine Sediments in Pacific Ocean

Many information about physical, chemical, nutritional parameters derive from oxygen and carbon isotope records, in particular from stack curves e.g., LR04 stack of Lisiecki and Raymo (2005) and that of Hoogakker et al. (2006). In a recent review Clark et al. (2006) discuss the meaning in sense of changes in deep-water temperature and ice volume contribution of the LR04 stack. The authors also report the possible mechanisms to cause the MPT and presenting evidences to support the hypothesis of the formation of thicker ice-sheets after the MPT respect to the pre-MPT time.

In the western equatorial Pacific e.g., Medina-Elizalde and Lea (2005; ODP Hole 806B) reveal the shift of SST oscillations from 41,000 to 100,000 yr, comparing their results with data from the eastern equatorial Pacific and investigating the nature and causes of the ciclycity observed. Always in the western equatorial Pacific (MD97-2140 core) de Garidel-Thoron et al. (2005) show SST stability during the Pleistocene and make a comparison with upwelling regions characterized by a long-term cooling trend. Winckler et al. (2005) reconstruct dust flux to central equatorial Pacific over the last 1 Myr from the study of the core TT013-114PC and suggest the dust levels to be higher in the last c. 560 kyr and minor during the MPT. A recent work (Hayward, et al., 2007) reveal that a peak of extinctions of benthic foraminiferal occurred during the MPT and propone possible mechanisms to justified these events.

Although a rich literature, as in this summary showed, is already available, a major number of paleoclimatic records from various latitudes and oceanographic settings is needed. In particular, with a multidisciplinary approach, the researches could be concern the causes of the MPT which are still unclear and objects of debate beyond provide a sufficient number of information for models.

REFERENCES

Bassinot FC, Beaufort L, Vincent E and Labeyrie L 1997. Changes in the dynamics of western equatorial Atlantic surface currents and biogenic productivity at the mid-Pleistocene revolution (□930 ka)In Shackleton NJ et al. (Eds) Proc. ODP, *Sci. Results* 154. College Station TX (ODP), 269-284.

Baumann K-H., Freitag T., 2004. - Pleistocene fluctuations in the northern Benguela Current system as revealed by coccolith assemblages. *Mar. Micropal., 52,* 195-215.

Berger, A., Jansen, E., 1994. – Mid-Pleistocene climate shift: the Nansen connection. In: Johannessen et al (Eds). The polar oceans and their role in shaping the global environment. *AGU Geophysical Monograph*, 85, 295-311.

Boyle EA 1981. Cadmium, zinc, copper, and barium in foraminifera tests. *Earth and Planetary Science Letters*, 53: 11-35.

Boyle EA 1988. Cadmium: Chemical tracer of deepwater paleoceanography. *Paleoceanography*, 3: 471-490.

Boyle, EA, Keigwin, LD, 1985. Comparison of Atlantic and Pacific paleochemical records for the last 215,000 years: changes in deep ocean circulation and chemical inventories. *Earth and Planetary Science Letters*, 76:,135-150.

Carter, L., McCave, I.N., 1994. Development of sediment drifts approaching an active plate margin under the SW Pacific Deep Western Boundary Current. *Paleoceanography*, 9, 1061-1085.

Carter, L., R.M Carter, I.N McCave, J. Gamble, Regional sediment recycling in the abyssal Southwest Pacific Ocean, *Geology,* 24, (1996a), 735-738.

Carter et al., 1996b Current controlled sediment deposition from the shelf to the deep ocean: the Cenozoic evolution of circulation through the SW Pacific gateway, 85, 438-451.

Carter L., and I.N. McCave, 1997, The sedimentary regime beneath the Deep Western Boundary Current inflow to the Southwest Pacific Ocean, *J. Sediment. Res*. 67, 1005-1017.

Carter, L., R.M. Carter, I.N. McCave, Evolution of the sedimentary system beneath the deep Pacific inflow off eastern New Zealand, *Mar. Geol.,* 205, (2004), 9-28.

Clark, P.U., Archer, D., Pollard, D., Blum, J. D., Rial, J. A., Brovkin, V., Mix, A. C., Pisias, N. G. Roy, M., 2006, The middle Pleistocene transition: characteristics. mechanisms, and implications for long-term changes in atmospheric pCO_2, *Quaternary Science Reviews*, 25, 3150-3184.

Clemens, SC, Murray, DW, Prell, WL, 1996. Nonstationary Phase of the Plio-Pleistocene Asian Monsoon. *Science*, 274:,943-948.

Crundwell M., Scott G., Naish T., and Carter L., Glacial–interglacial ocean climate variability from planktonic foraminifera during the Mid-Pleistocene transition in the temperate Southwest Pacific, ODP Site 1123, *Palaeogeography, Palaeoclimatology, Palaeoecology*, in press.

de Garidel-Thoron T., Rosenthal Y., Bassinot F., et al., 2005, Stable sea surface temperatures in the western Pacific warm pool over the past 1.75 million years, *Nature,* 433, 294-298.

de Kaenel, E., Siesser, W.G., Murat, A., 1999. Pleistocene calcareous nannofossil biostratigraphy and western Mediterranean sapropels, sites 974 to 977 and 979. In: Comas, Z.R., Klaus, A. (Eds.). Proceeding ODP Leg 161. *Scientific Results*, 161, 159-183.

Dwyer, G.S., Cronin T.M., Baker P.A., Raymo M.E., Buzas J.S., Correge T., 1995. North Atlantic deep water temperature change during late Pliocene and late Quaternary climatic cycles. *Science,* 270, 1347-1351.

Fenner, J., L., Carter, and R, Stewart, 1992, Late Quaternary paleoclimatic and paleoceanographic change over northern Chatham Rise, New Zealand, 108, 383-404.

Fenner, J., and Di Stefano, A., 2004, Late Quaternari oceanic fronts along Chatham Rise indicated by phytoplankton assemblages, and refined calcareous nannofossil stratigraphy for the mid-latitude SW Pacific, *Marine Geology*, 205, 59-86.

Flores, J.A., Sierro, F.J., 2007. Pronounced mid-Pleistocene southward shift of the Polar Front in the Atlantic sector of the Southern Ocean. *Deep-Sea Research II*, 54, 2432–2442.

Flores, J.A., Gersonde, R. and Sierro, F.J., 1999. - Pleistocene fluctuations in the Agulhas Current Retroflection based on the calcareous plankton record. *Marine Micropaleontology,* 37, 1-22.

Flores, J.A., Gersonde, R., Sierro, F.J., Niebler, H.S., 2000. – Southern Pleistocene calcareous nannofossil events. Calibrations with isotope and geomagnetic stratigraphies. *Marine Micropaleontology,* 40, 377-402.

Hall, I.R., McCave, I.N., Shackleton, N.J., Weedon, G.P., Harris, A.H., 2001. Intensified deep Pacific inflow and ventilation in Pleistocene glacial times. *Nature*, 412, 809-812.

Hayward B.W., 2001. Global deep-sea extinctions during the Pleistocene ice-ages. *Geology,* 29, 599-602.

Hayward, 2002 B.W. Hayward, Late Pliocene to middle Pleistocene extinctions of deep-sea benthic foraminifera ("*Stilostomella* extinction") in the Southwest Pacific, *J. Foraminiferal Res.* 32 (2002), pp. 274–306.

Hayward, B.W., Kawagata, S., Grenfell H., R., Sabaa, A., T., 2007, Last global extinction in the deep sea during the mid-Pleistocene climate transition, PA3103, doi:10.1029/2007PA001424.

Hoogakker, B.A.A., Rohling, E.J., Palmer M.R., Tyrrell T., and Rothwell R.G., 2006, Underlying causes for long-term global ocean $\delta^{13}C$ fluctuations over the last 1.20 Myr, 248, 15-29.

Huybers, P., 2007, Glacial variability over the last two million years: an extended depth-derived agemodel, continuous obliquity pacing, and the Pleistocene progression, *Quaternary Science Reviews*, 26, 37-55.

Kawagata S., Hayward B. W., Grenfell H. R., Sabaa A. 2005. Mid-Pleistocene extinction of deep-sea foraminifera in the North Atlantic Gateway (ODP sites 980 and 982) *Palaeogeography, Palaeoclimatology, Palaeoecology*, 221, 167-291.

Kawagata S., Hayward B. W., Gupta A. K. 2006. Benthic foraminiferal extinctions linked to late Pliocene–Pleistocene deep-sea circulation changes in the northern Indian Ocean (ODP Sites 722 and 758) *Marine Micropaleontology*, 58, 219-242.

Kitamura, A, Kawagoe, T, 2006. Eustatic sea-level change at the Mid-Pleistocene climate transition: new evidence from the shallow-marine sediment record of Japan. *Quaternary Science Reviews,* 25: 323-335.

King, A. L., and W. R. Howard, 2000, Middle Pleistocene sea-surface temperature change in the southwest Pacific Ocean on orbital and suborbital time scales, *Geology*, 28, 659-662.

Kucera, M, Kennett, JP, 2002. Causes and consequences of a middle Pleistocene origin of the modern planktonic foraminifer *Neogloboquadrina pachyderma* sinistral. *Geology*, 30(6): 539-542

Lea, DW, Pak, DK, Spero, HJ, 2000. Climate Impact of Late Quaternary Equatorial Pacific Sea Surface Temperature Variations. *Science,* 289: 1719-1724.

Lisiecki, L.E., Raymo, M.E., 2005. A Pliocene-Pleistocene stack of 57 globally distributed benthic $\delta^{18}O$ records. *Paleoceanography,* 20, doi:10.1029/2004PA001071.

Lupi, C., 2007. – Il nannoplancton calcareo e le sue relazioni con il sistema climatico negli oceani del Quaternario. Tesi di Dottorato in Scienze della Terra, XIX Ciclo, Università degli Studi di Pavia. PhD Thesis.

Maiorano, P., Marino, M. 2004. – Calcareous nannofossil bioevents and environmental control on temporal and spatial patterns at the early-middle Pleistocene. *Marine Micropaleontology*, 53, 405-422.

Marchitto, TM, Oppo, DW, Curry, WB, 2002. Paired benthic foraminiferal Cd/Ca and Zn/Ca evidence for a greatly increased presence of Southern Ocean Water in the glacial North Atlantic Paleoceanography, 17 doi:10.1029/2000PA000598.

McCave, I.N., Carter, L., 1997. Recent sedimentation beneath the Deep Western Boundary Current off northern New Zealand. *Deep-Sea Res. I,* 44, 1203-1237.

Medina-Elizalde M. and Lea D.V., 2005, *The Mid-Pleistocene Transition in the tropical Pacific,* 310, 1009-1012.

Nelson C. S., Cooke, P.J., Hendy C. H., Cuthbertson, A. M., 1993, Oceanographic and climatic changes over the past 160 000 years at Deep Sea Drilling Project Site 594 off southeastern New Zealand, Southwest Pacific Ocean, *Paleoceanography*, 8, 435-458.

Nelson C. S., Hendy I.L., Neil H.L., Hendy C. H., Weaver, P.P.E., 2000, Last glacial jetting of cold waters through the Subtropical Convergence zone in the Southwest Pacific off eastern New Zealand, and some geological implications, 156, 103-121.

Nodder S. D, Pilditch, C. A., Probert, P. K., and Hall J., A., 2003, Variability in benthic biomass and activity beneath the Subtropical Front, Chatham Rise, SW Pacific Ocean, *Deep-Sea Research I*, 50, 959-985.

Prell W.L., 1982. Oxygen and carbon isotope stratigraphy for the Quaternary of the hole 502B : evidence for two modes of isotopic variability. *Initial reports DSDP*, 68, 455-464.

Raffi, I., 2002. Revision of the early-middle Pleistocene calcareous nannofossil biochronology (1.75-0.85 Ma) *Marine Micropaleontology*, 45, 25-55.

Raffi I., Backmann J., Fornaciari E., Palike H., Rio D., Lourens L., Hilgen F., 2006. A review of calcareous nannofossil astrobiochronology encompassing the past 25 million years. *Quaternary Science Review*, doi: 10.1016/j.quascirev.2006.07.007.

Raymo, ME., 1997. The timing of major climate terminations. *Paleoceanography*, 12: 577-585.

Ruddiman W., Raymo M.E., Martinson D., Clement B., Backman J., 1989. Pleistocene evolution: Northern Hemisphere ice sheets and North Atlantic Ocean. *Paleoceanography*, 4, 353-412.

Schrag D.P., Hampt G., Murray D.W.S., 1996. Pore fluid constraints on the temperature and oxygen isotopic composition of the glacial ocean. *Science,* 272, 1930-1932.

Schrag D.P., Adkins J.F., McIntyre K., Alexander J.L., Hodell D.A. Charles C.D., McManus J.F., 2002. The oxygen isotopic composition of seawater during the Last Glacial Maximum. *Quaternary Science Review*, 21, 331-342.

Shackleton, N.J., Berger, A., Peltier, W.R., 1990. An alternative astronomical calibration of the Lower Pleistocene timescale based on ODP Site 677. *Transactions of the Royal Society of Edinburgh: Earth Sciences*, 81, 251–261.

Sun Y., Clemens S.C., An Z., Yu Z., 2006. Astronomical timescale and paleoclimatic implicatins of stacked 3.6-Myr monsoon records from the Chinese Loess Plateau. *Quaternary Science Reviews*, 25, 33-48.

Teitler, L, Kupp, G. Warnke, D, Burckle L., 2007. Evidence for a long warm interglacial during Marine Isotope Stage 31: Comparison of two studies at proximal and distal marine sites in the Southern. US Geological Survey and The National Academies; USGS OF-2007-1047, Extended Abstract 013.

Tziperman, E, Gildor, H, 2003. On the mid-Pleistocene transition to 100 kyr glacial cycle and the asymmetry between glaciation and deglaciation times. *Paleoceanography*, 18, 1001, doi:10.1029/2001PA000627.

Xu J., Wang P., Huang B., Li Q., Jian Z., 2005. Response of planktonic foraminifera to glacial cycles: Mid-Pleistocene changes in the southern South China Sea. *Marine Micropaleontology,* 54, 89-105.

Venz, KA, Hodell, DA., 2002. Plio-Pleistocene record of deep-water circulation in the Southern Ocean from ODP Leg 177, Site 1090. *Palaeogeography, Palaeoclimatology, Palaeoecology*., 182: 197-220.

Venuti A., Florindo F., Michel E., Hall I. R., 2007, Magnetic proxy for the Deep (Pacific) Western Boundary Current variability across the Mid-Pleistocene climate transition, *Earth and Planetary Science Letters*, 259, 107-118.

Villa G, Lupi C, Cobianchi M, Florindo F and Peker SF 2007. A Pleistocene Warming Event at 1 Ma in Prydz Bay, East Antarctica: Evidence from ODP Site 1165 U.S. Geological Survey and The National Academies; USGS OF 2007–1047, Extended Abstract 205.

Wang, P., Tian, J, Cheng, X., Liu, C., Xu, J., 2003 – Carbon reservoir changes preceded major ice-sheet expansion at the mid-Brunhes event. *Geological Society of America* 31, 239-242.

Whitworth, T., Warren, B.A., Nowlin, W.D., Rutz, S.B., Pillsbury, R.D., Moore, M.I., 1999. On the deep western-boundary current in the Southwest Pacific Basin. *Prog. Oceanogr.*, 43, 1-54.

Winckler G., Anderson, R., F., and Schlosser, P, 2005, Equatorial Pacific productivity and dust flux during the mid-Pleistocene climate transition, Paleoceanography, 20, PA4025, doi:10.1029/2005PA001177.

Zheng F., Quianyu L., Baohua L., Muhong C., Xia T., Jun T., Zhimin J., 2005. A millennian scale planktonic foraminifer record of the mid-Pleistocene climate transition from the northern South China Sea.. *Palaeogeography, Palaeoclimatology, Palaeoecology*, 223, 349-363.

RESEARCH AND REVIEW STUDIES

In: The Pacific and Arctic Oceans
Editor: Kallen B. Tewles

ISBN: 978-1-60692-010-7

Chapter 1

BENTHIC AND PELAGIC STUDIES IN THE COASTAL MEXICAN PACIFIC

V. Díaz-Castañeda, L. Ladah, L. Calderon-Aguilera and J. Farber-Lorda

División de Oceanología, CICESE, Km 107 Carretera Tijuana-Ensenada, Ap. Postal 2732, Ensenada, Baja California. Mexico

INTRODUCTION

Mexico is one of the 12 megadiverse countries in the world (Mexico, Colombia, Ecuador, Peru, Brazil, Zaire, Madagascar, China, India, Malaysia, Indonesia, Australia). These countries have greater than 70% of the world's biodiversity (McNeely et al., 1990). Mexico has a spectacular 10,000 km of coastline (Farreras, 2006) which is known for its kelp forests, coral reefs, mangroves and coastal lagoons, generally characterized by high productivity, low anthropogenic impact, and high biodiversity. Much of these ecosystems are understudied and it is important to establish baseline data in coastal areas of Mexico as they are threatened by potential (or current) development and pollution which will certainly change the natural fauna and flora of these important areas. The lagoons and coasts of California, USA, are considered endangered habitats due to the loss of 90% of their original area through urbanization (Talley and Ibarra-Obando, 2000). However, neighboring Baja California shows comparatively little anthropogenic impacts to date, yet is faced with the looming threat of rapid and unsustainable development. In this context, an analysis of faunal and environmental conditions of the coastal areas of Mexico is essential for control studies and to establish a baseline.

OCEANOGRAPHIC FEATURES OF THE MEXICAN PACIFIC

The Mexican Pacific coast exists in a climatic transition zone from temperate to sub-tropical and tropical waters, from the California Current to the Western Mexican Current and the Costa Rica Coastal Current (Figure 1). Upwelling exists in various areas along the

Mexican Pacific, mainly along the Baja California peninsula and in the Tehuantepec regions (Figure 2). Historically, most studies have focused on the temperate and sub-tropical coasts of Mexico, due to the geographic situation of most of the research centers in Mexico. Recently, however, more emphasis is being focused on tropical regions as well.

The oceanography of the Pacific Ocean of the Baja California peninsula is dominated by the southward flowing California Current (CC) and coastal wind-induced upwelling of cold nutrient-rich water from below the thermocline (Figure 2). Dominant upwelling areas include the prominent outcroppings or points south of Ensenada, some of them known as the 'seven sisters' with fairly limited road access, down to the large conspicuous point of Punta Eugenia. The furthest south traditional upwelling site on the Baja California coast, Bahía Magdalena, represents the southern-most record of coastal shallow kelp forests, beyond which wind-induced upwelling becomes much less relevant and the system transitions quickly to a more subtropical/tropical flora and fauna. As upwelling becomes less prevalent, surface waters remain warm and tend not to be replenished with cool nutrient-rich water masses, and therefore often become oligotrophic. The warming of waters as one progresses further south along the coast of the Baja California peninsula and the reduced input of nutrients into these surface waters by coastal upwelling, results in greater stress on temperate organisms that depend on the process of coastal upwelling, which is typical in Eastern Boundary currents, thereby resulting in distributional limits of many temperate organisms on this coast and a switch from temperate benthic kelp forests to tropical coral reefs.

The CC (California Current) is an important input and boundary condition for the Baja California region (Lavin et al., 2006) in the Northern Mexican Pacific. El Niño Southern Oscillation (ENSO) is an important phenomenon throughout the world and particularly in this area. Its effects on marine ecosystems and organisms may go beyond temperature changes. Marine invertebrates have complex life cycles in which certain life stages, and hence the dynamics of entire populations, are at the mercy of various physical processes acting within the ocean-atmosphere system (Bakun 1996; Escribano et al., 2004). Higher temperatures and lower nutrient concentrations can result in widespread mortality of macroalgae, corals, seagrasses and macrobenthic organisms in this current.

In the Eastern Tropical Pacific, the dominant currents are: the Costa Rican Coastal Current that flows poleward, but loses its identity and merges with the Tehuantepec Bowl and the North Equatorial Countercurrent. The other current in the region is the West Mexican Current which also flows poleward but is a subsurface current and flows closer to the coast. Kessler (2006) also propose the Tehuantepec Bowl, which acts like a Gyre that takes some of the Southern California Current and the Costa Rica Coastal Current and flows to the Central Pacific (Figure 1).

The eastern tropical region of the Pacific has been considered a biogeographically distinct province (Ebeling, 1967; McGowan, 1971, 1974, 1986; Briggs, 1975; Vermeij, 1978; Hastings, 2000) In this area there exists the Oxygen Minimum Zone (OMZ), which shows a very shallow, almost anoxic, layer up to 1,000 m thick and occurs at shallower depths than other low-oxygen areas like the Humboldt Current and the Arabian Sea (Helly and Levin, 2004; Fernandez-Álamo and Färber-Lorda, 2006). In most of the low oxygen areas, oxygen is <1.5 ml.l-1 or of <20% saturation, whereas in the eastern tropical Pacific, oxygen is 0.5 ml.l-1 or 7.5% of saturation (Kamykowski and Zentara, 1990; Helly and Levin, 2004). It is evident that the Mexican tropical Pacific has a very shallow OMZ.

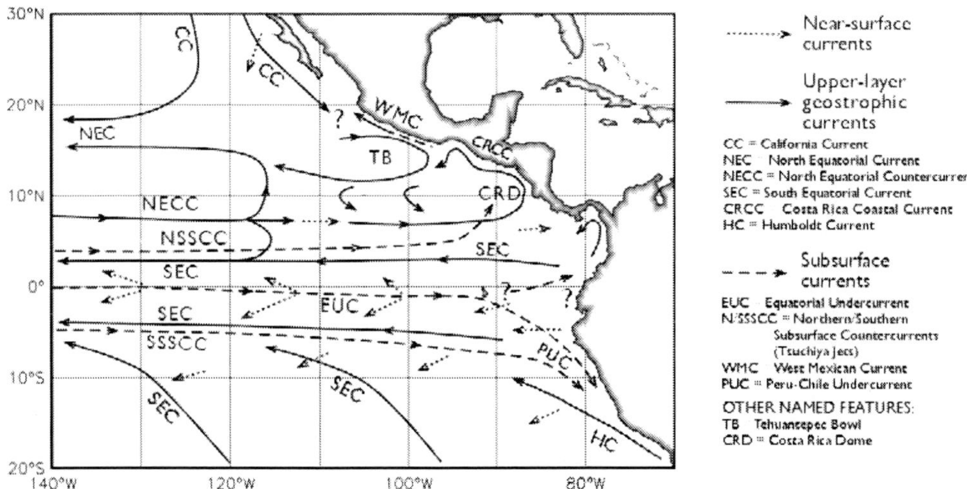

Figure 1. Three-dimensional circulation in the eastern tropical Pacific, as proposed by Kessler (2006) with the author's permission. Question marks indicate regions where the interconnections are still not known.

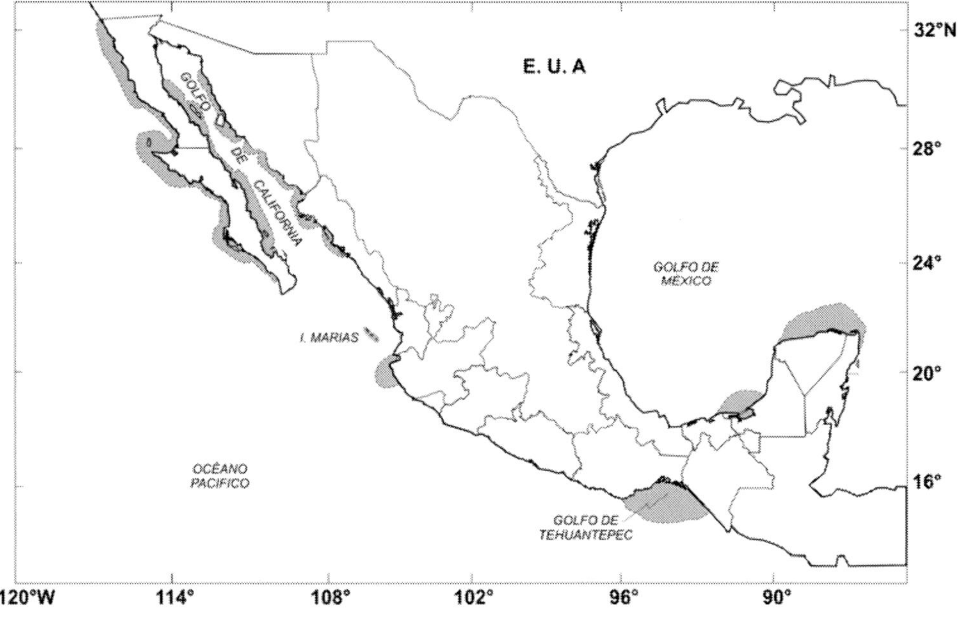

Figure 2. Regions were upwelling takes place along the coast of Mexico.

In the eastern tropical Pacific, the mesoscale features are mostly associated with the wind jets blowing through three gaps in México and the Central American cordillera: at the Tehuantepec isthmus in Mexico, the Gulf of Papagayo region of Nicaragua, and the Panama Canal (Amador, Alfaro, Lizano and Magaña, 2006). This produces jets that force strong mixing and upwelling (Kessler, 2006; Willett et al., 2006) and is reflected in high biological productivity in the area (Färber-Lorda, Lavín, Zapatero and Robles, 1994; Färber-Lorda

Lavín and Guerrero,2004b; Gonzalez-Silvera, Santamaría-del-Angel, Millán-Nuñez and Manzo-Monroy, 2004; Pennington *et al.*, 2006).

Indicator Communities

In this chapter, we will review four important indicator communities of the Mexican Pacific. We will begin with the temperate kelp forest and rocky shore communities of temperate Baja California and Baja California Sur. We will then move to macrofaunal benthic communities with a principal focus on polychaetes. After that we will proceed to review corals in the Mexican Pacific and then finish with zooplankton in the Eastern Tropical Pacific. All of these communities are key species in their ecosystems and can be early indicators for contamination and climate change.

Kelp Forest and Rocky Shore Communities

Background

Kelp Forests

The kelp forest ecosystem in Mexico is an ecologically and socio-economically valuable resource, which is rich in biodiversity, ecological services, end-users, and fisheries, and in Mexico only occurs along the Baja California peninsula. The kelp forest ecosystem provides food, habitat, and refuge for many coastal marine organisms and is, unfortunately, highly susceptible to climate change and ocean warming. In particular, it has been somewhat perplexing to understand both how temperate kelp populations can survive in central and southern Baja California where water temperatures are seasonally warm (Hernandez Carmona *et al.* 2001, Ladah and Zertuche 2004) and how the populations can recover rapidly after severe disturbance caused by El Niño events (Ladah *et al.* 1999, Edwards 2004).

Conventionally, delivery of nutrients to kelp forests has been ascribed to wind-induced upwelling of cool nutrient-rich water from below the thermocline. Upwelling accounts for a large part of annual inorganic nitrogen transport to kelp forests in California (approximately 70%, McPhee-Shaw *et al.* 2007). However, there exist other mechanisms of nutrient delivery to kelp forests. These include coastal eddies (Bassin et al. 2005), and internal waves (Zimmerman and Kremer 1984, McPhee-Shaw *et al.* 2007). An internal wave is simply a wave form that develops within the water column between layers of water of different densities. Internal waves can cause a 10 fold increase in available nitrate and have been shown to cause rapid changes in temperature associated with vertical and horizontal displacements of water masses (and associated nutrients) in temperate settings (e.g. Sandstrom and Elliott 1984, Pineda 1991, Witman *et al.* 1993).

Linking nearshore physics to kelp forest biology, with a particular emphasis on nutrient transport mechanisms, is of particular interest in the kelp forest ecosystem and has not been studied in this important biogeographic transition zone. In particular, in an area where traditional upwelling may become less relevant, other nutrient delivery mechanisms such as internal waves may play an important role in maintaining kelp populations, especially near their southern limit where upwelling is often intermittent or even absent in some years. Because some of the alternative nutrient delivery mechanisms besides upwelling depend strongly on coastal water column stratification, understanding how stratification will change in light of global warming is an important consideration for the survival of the kelp populations in the southern limit region.

Rocky Shore Benthic Invertebrates

Many benthic invertebrates have a pelagic larval phase lasting from days to months in the open ocean. These larvae require some type of physical advective process to move them onshore and eventually reach adult habitats where they can complete their life cycle (Shanks 1983; Farrel et al. 1991; Pineda 1991; Le Fèvre and Bourget 1992; Shanks 1995; Pineda 1999; Shanks et al. 2000). Transport mechanisms include wind-induced upwelling and relaxation (Farrel et al. 1991; Wing et al. 1995a; Wing et al. 1995b; Shanks et al. 2000) and internal waves (Norris 1963, Pineda 1991, 1994a,b). Physical processes mediating the transport of invertebrate planktonic larvae have been shown to affect the timing of settlement (Queiroga and Blanton 2005, Ladah et al. 2005), however other physical and biological processes may shape the spatial pattern of larval delivery and settlement (Gaines et al. 1985), resulting in small-scale spatial heterogeneity (e.g. Ladah et al. 2005). Distribution of the nearshore pool of larvae also vary spatially and temporally, due to variability in the timing and intensity of adult reproduction (Pineda 2000, Macho et al. 2005). These processes may work in concert with the vertical distribution and behavioral traits of larvae to result in heterogeneous settlement along a shore.

Understanding how pelagic larvae of benthic invertebrates are transported to shore is important for the replenishment and maintenance of many invertebrate populations near their southern limit in this biogeographic transition zone. For example, both purple and red sea urchins and various species of abalone are harvested from this coast, yet little is known of their larval transport. Easy to study proxy species, such as barnacles, can shed light on invertebrate larval transport mechanisms, and on how delivery of larvae differs spatially and temporally. Also because larval transport only seems to occur during specific peaks dependent on tides and coastal water column stratification, how these forcing factors may change over time becomes a relevant question. Without the replenishment of larvae from these physical transport mechanisms, populations near their southern limit may also become extinct.

General Methods

Physical Oceanography

In order to link biological parameters with physical conditions, a general understanding of the coastal water column just offshore of the population in question is necessary. Important factors, such as surface and bottom water temperature, depth of the thermocline, water column stratification, and how these factors change over time, are crucial to understanding coastal zone fluxes. Coastal surface and subsurface currents are also an interesting facet to understanding transport, but are not always necessary, as much can be deduced from high frequency (every 5 to 15 minutes) water column measurements.

Water column temperature measurements can be acquired in two ways. One way is using CTD (conductivity, temperature, depth) stations/transects taken with high frequency (for example, daily, for a week, every few hours). The easier method is to use one (or many) thermistor chain(s) anchored offshore of the study site in shallow (<30m depth) water, with recording thermographs (Hobo loggers from Onset Computer Corp. are a reliable and inexpensive option) set to read at high frequency (every 1 to 15 minutes), attached every few meters along the chain. It is most important to have at least one thermsitor above and one below the thermocline that will be moving with the internal tide. If possible, it is best to combine both methods to have both temporal and spatial understanding of the water column just offshore.

Currents can be measured also in two ways, and both are interesting for biological-physical coupling and transport studies, and although providing additional important information, these methods are more expensive than using thermistor strings. The first method, using surface satellite drifters, is useful for understanding the movement of surface currents. The second method, most important for water column currents, is to use a bottom mounted upwards-looking ADCP (Acoustic Doppler Current Profiler) measuring every minute if possible in bins of 1 meter. A downwards-looking boat mounted ADCP is another option, which can then be used to take transects similar to the CTD transects in the previous section. Again if both can be used, a better understanding of the spatial and temporal variability of the water column can be acquired.

Biological Oceanography

If nutrients are part of the proposed study question, then nutrient concentrations (usually nitrate, nitrite and sometimes ammonia are important) need to be acquired, preferably at the same time and depth of temperature measurements. This serves ultimately to propose a temperature-nutrient relationship. Once reliably established, temperature can then (cautiously) be used as a nutrient proxy. CTD casts using Niskin bottles are the traditional oceanographic method for collecting nutrients. Seawater is immediately filtered with a syringe and GFF (glass fiber filters), frozen, and then analyzed on an autoanalyzer or using the traditional cadmium reduction method. A new and expensive method using an underway nitrate detector called ISUS takes advantage of a UV nitrate sensor and can be placed on a CTD to get nitrate data in real time as well. Please see Ladah (2003) or Zimmerman and Kremer (1984) for a detailed review of establishing the temperature-nutrient relationship, and see Zimmerman and Robertson (1985) for how this can vary between years.

In the case of nutrient delivery to nearshore macroalgae or kelps, it is important to measure nutrient content, uptake kinetics, and enzymatic activity at a high frequency in the seaweed tissue at the same depths that temperature (representative of deep water high frequency fluxes) and/or currents (transport) are measured. The methods to get at this information range from extremely complex biochemical methods, to the simple harvesting, drying, and processing of algal tissue for Carbon, Nitrogen, and Hydrogen content. Nitrogen content using a CNH analyzer will give total nitrogen content, including still labile nitrate. Understanding basic biological parameters, such as population dynamics (density, recruitment, survivorship, reproductive periods, etc.), is important background information necessary to understand the dependence of the population on physical conditions. Detailed studies on kelp demography in Baja California can be found in Ladah *et al.* (1999), Ladah and Zertuche (2004), as well as references therein.

For measurement of invertebrate larvae, there are also two approaches. Larvae can be collected in the water column as well as on the shore. High frequency transects of water column plankton samples using surface and vertically integrated net tows to identify the types of larvae transported by different mechanisms would be important. These samples can then be stored in formalin and analyzed/enumerated using a microscope at a later date. Onshore larval settlement can be measured either by using PVC tubes or plexiglass plates for different barnacles, attaching them with a removeable screw in the intertidal or, for urchin, seastar, or crab larvae, attaching SOS tuffy pads (kitchen cleaning sponges) in the intertidal (see Pineda 1999 and Sanford and Menge 2007 for methods). These can then be collected every so often (daily resolution is best, depending on the frequency of the physical mechanism you wish to correlate with), enumerated with a dissecting scope, cleaned and replaced. Many larvae are easy to differentiate (barnacles), others require some training, but once the eye becomes adjusted to recognizing specific taxa, this method is quick and easy and gives high resolution results.

Correlating Physics and Biology

One of the toughest problems that biologists face when attempting to work with physical oceanographers is that the acquisition of physical data at a high frequency is often only limited by the resolution of the equipment or battery life and storage space. On the other hand, biological data is usually limited by human energy and by microscope time. For example, a CTD can sample salinity and temperature every second for an entire cruise, whereas a biological net tow or nutrient cast can be performed at best possibly every half hour. Because of this issue, biological and physical data are usually collected in different time frames. This poses the first challenge in linking biological and physical data sets, which can however be overcome through integration of physical data over periods of time of biological samples.

The second, and potentially larger challenge to linking biological and physical data sets, is that correlation does not imply causation. Therefore, unless a specific physical mechanism is at least reasonably well understood, as in the case of internal wave theory, it becomes difficult to attribute certain correlations to cause-effect factors.

Results

Temperature

Water column temperatures taken over long periods of time can identify periods of bottom cooling and surface warming, both of which are important for linking the coastal biology of temperature-sensitive organisms at their southern limit to their surrounding physical conditions. In general, strong stratification and surface warming occurs in most sites in Baja California in late summer and early fall (see Figure 3a for an example). Many sites also show the typical pattern of bottom cooling in spring and early summer as shown in Figure 3a. Temperature records in general show many clear examples of cooler than ambient water intruding into the bottom third of the water column at all sites for short periods of time (Figure 3b) often oscillating with the warm phase of an internal wave. This can cause pulses of cold water which intrude into shallow sites for hours at a time (Figure 3b, see Panel A, 21 to 24 hrs). If linked with nutrient data from a previously established temperature-nutrient relationship, these long term temperature time series, which are relatively cheap and easy to acquire, can give valuable information on nutrient availability at a particular site when the temperature-nutrient relationship is applied with caution (it can change slightly during different years, see Zimmerman and Robertson 1985). Many times, water temperatures near the thermocline can change 5 degrees C in less than an hour when the internal wave passes (Figure 3b, panel A, 21 hr), having obvious implications for physiology and nutrient transport. All temperature time series this author has measured so far across Baja California are highly variable at daily and higher frequencies suggesting strong internal wave activity, especially during periods of extreme tidal forcing and strong stratification.

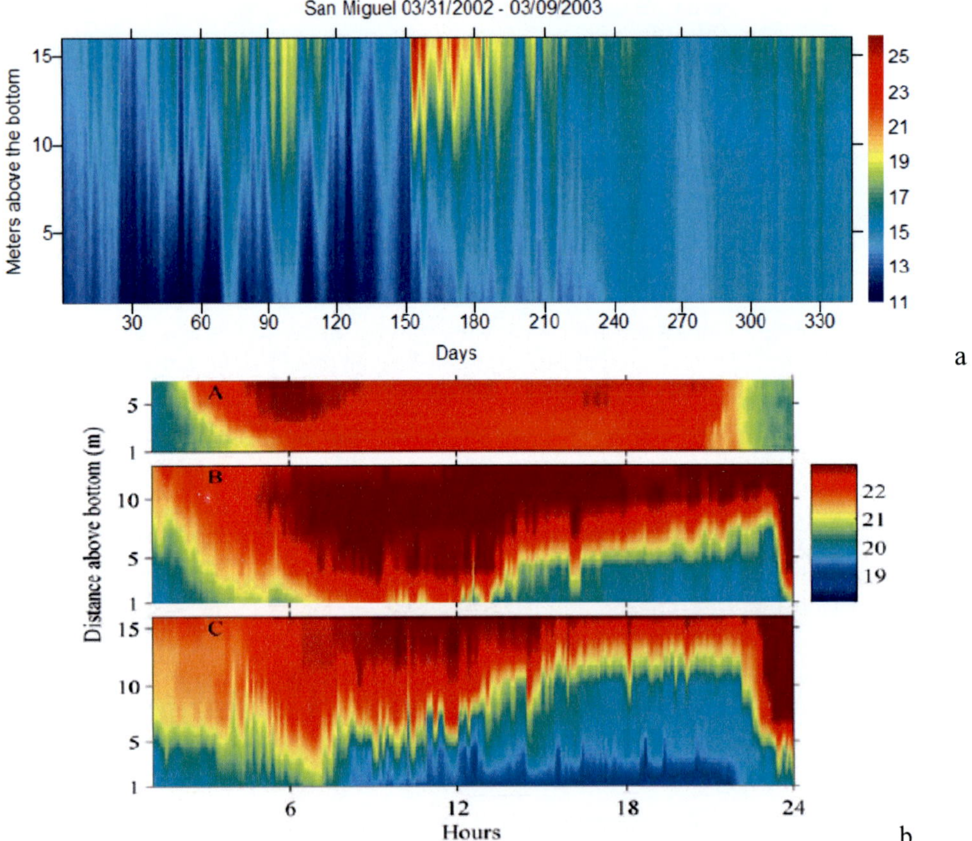

Figure 3. Examples of water column temperature time series. 3-a shows long term time series within 1km from shore for Northern Baja California using Onset Hobo temperature loggers set to read every minute at every 3m depth. Note strong stratification in the month of September 2002. 3-b, in which panels A, B, and C refers to 8, 15, and 20m depth in a transect offshore, all within 2km of the coastline, shows a zoom of a 24 hr. period in August 19, 2006 from Southern Baja California, very near the kelp forest southern limit and shows a strong internal tide with warm water reaching depth and colder water being pulsed into the shallow site where the kelp bed thrives off this strong high frequency pulsing lasting only a few hours. Right legends show temperature in Celsius. See Ladah in review and Ladah and Leichter in review for details.

Currents

Current and flow data for the water column in various sites of Baja California show short periods of very intense sign oscillations, meaning that currents change from onshore to offshore, and from up to down, very quickly, various times, in a very short period of time (changing every 15 minutes or so within a hour) (Figure 4, middle panels). This change of sign is often representative of passing internal waves, and occurred in this N. Baja Caliornia site shortly after high tide, most likely due to the extreme forcing of the tides changing from low to high, and the time lag for the internal waves to propagate to the nearshore from where they were created offshore at the shelf break.

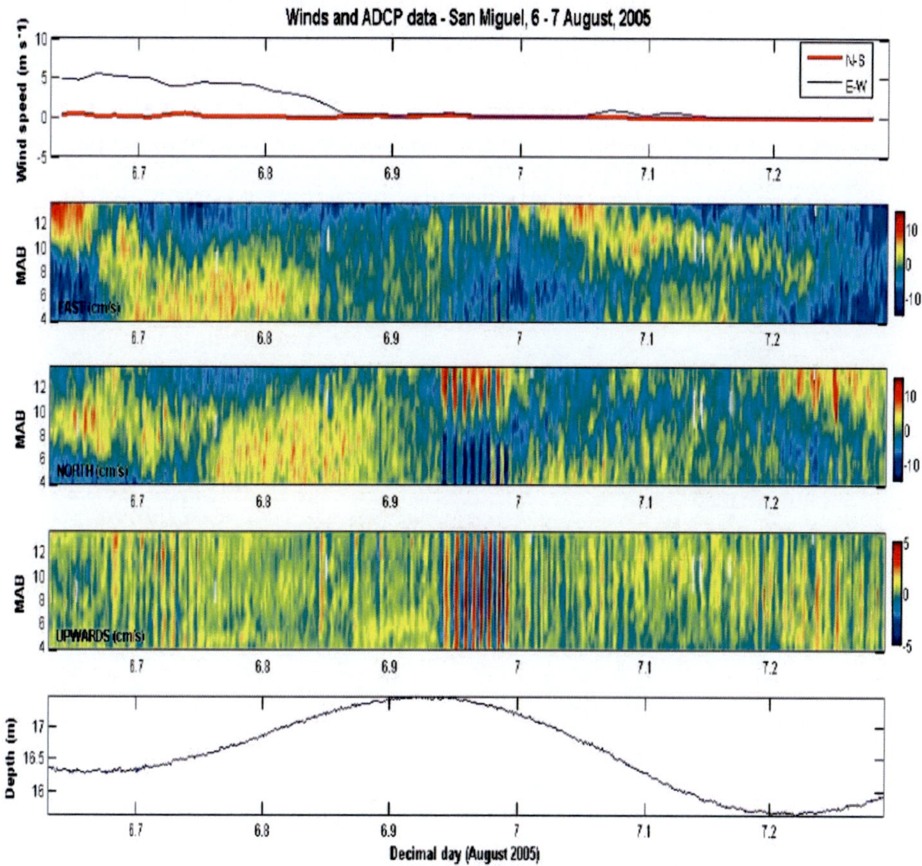

Figure 4. Winds and Raw (unprocessed) ADCP data from Northern Baja California for less than 1 day on August 2006 (zoom in on decimal day 6.7 to 7.2, Aug 6, 2006, 4pm to Aug 7, 2006, 4am), showing current data in the N-S, E-W, and vertical direction. Note strong change of sign in N-S and vertical currents suggesting passing high frequency internal waves lasting less than 15 minutes. Right axis is velocity in cm/s. See Ladah *et al.* 2005, and Ladah and Tapia in review for details.

Temperature-Nutrient Relationship

Temperature data combined with nutrients can be used to provide a temperature proxy for nutrients. However, across Baja California, this proxy changes drastically with latitude; most probably due to transport of different water masses and therefore must be established for a particular site and time (see Ladah 2003). For example, in Northern Baja California such as in Ensenada and Punta Baja, the temperature-nutrient relationship behaves in a similar fashion to that established for California (see Zimmerman and Kremer 1984) with a lack of nutrients in water warmer than 15C. However, as one moves further south to sites such as Abreojos or San Juanico, the temperature-nitrate relationship changes. Nitrate becomes available in much warmer waters (up to 19C) (Figure 5) (see Ladah 2003 for a full analysis).

Biological-Physical Coupling

Once profiles of water column temperatures and currents are available, the physical data can be correlated with biological parameters taken at the same time. For example, we can correlate the number of barnacle cyprid larvae settled on collection plates with water column temperature offshore, keeping in mind that stratification is a requirement for internal wave activity. A strong positive correlation is shown between stratification between 7 and 10 meters above the bottom, where the thermocline was at this time, and the number of settlers at a single site in Northern Baja California (Figure 6). This pattern was found to hold for many sites and years (Ladah et al. 2005).

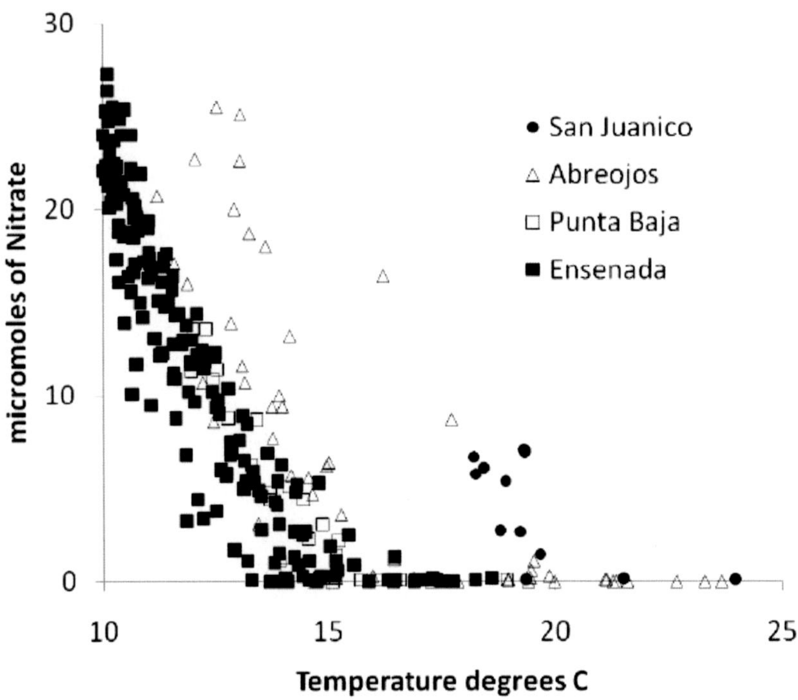

Figure 5. Relationship between temperature and nitrate in coastal waters in less than 50m depth at various locations along the Baja California Peninsula, showing that in sites further south (such as Abreojos and San Juanico), there is nutrient enrichment in warmer water than further north in the typical upwelling areas of Ensenada and Punta Banda. See Ladah 2003, and Ladah and Leichter in review for an explanation of sites.

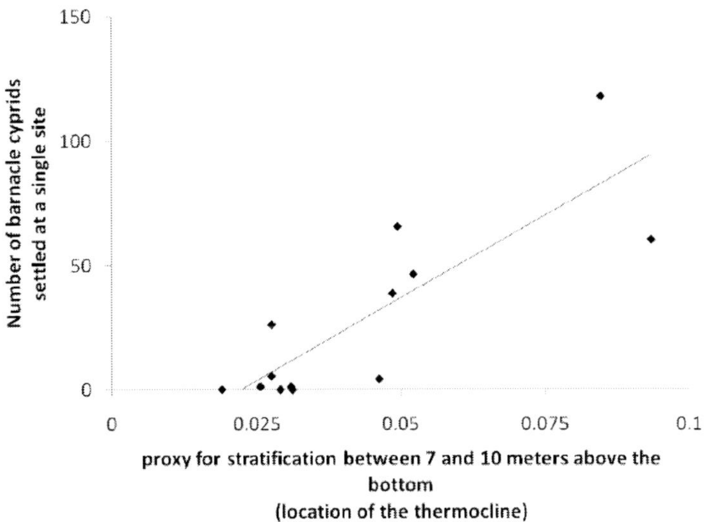

Figure 6. Correlation between a single monitoring site for 11 days at San Miguel for barnacle larvae cyprid settlement and offshore water column characteristics. In this case the Brunst Vaisella frequency between 7 and 10m, generally where the thermocline is located was integrated over the times of larval collector deployments to then be able to correlate data. Note that not only timing but also intensity of larval settlement as well correlates strongly with offshore water column internal motions (movement up and down of the thermocline). $R=0.834$, $p<0.001$. See Ladah and Tapia in review for a detailed explanation.

Relevance

The effects of short term perturbations, or long-term environmental shifts (e.g., climate change) on ecosystem functioning, are likely to be magnified in biogeographic transition zones. Linking the biology and the physics of the coastal temperate zone in Mexico, where many species encounter their southern limits, has relevance for local management of the kelp forests, for the fished organisms that depends on kelp forests for food, and for understanding invertebrate larval transport for recruitment in fisheries. Kelps play a key role in providing habitat, food, and refuge for many coastal marine organisms, some of which are commercially exploited, such as abalone, urchin, snails, and lobster, thus having great ecological and social importance. Scientific results will provide a greater understanding of connectivity and sustainable management of these ecosystems.

Future Research

There exist many opportunities for future research in the field of biological-physical coupling of high frequency mechanisms. Few studies actually look at specific mechanisms for transport of larvae and nutrients, less so in the high frequency range in the ocean, because these changes are difficult to study and even harder to document well. Many approaches require diving in rough, shallow, coastal conditions in temperate areas, with dangerous conditions within dense kelp forests, which results in limited bottom time. Intertidal work with invertebrate larvae and intertidal kelps requires low tide periods, which only last a matter of hours on certain days of the month. And while the physical measurements are becoming easier and easier with the development of new technology for accuracy and storage of data, the biological measurements are still tedious, labor intensive and require many people concurrently working across the shore.

A major gap in data is that continuous bottom temperature measurements are often unavailable for many of the remote areas of the Baja California coast, and even if available, have rarely been collected at high frequencies. Increased collection of time series data and coastal monitoring in this region would be extremely valuable for understanding the temporal and spatial patterns of the physical mechanisms on the coast. Only then can the physics be linked with the biology.

Combining biochemical and physiological studies on organisms and how they respond to the temperature and nutrient changes that result when internal wave activity occurs would be of great interest as well. For example, photosynthetic physiology and nutrient uptake rates could be measured *in situ* at the same time and could yield important results for the understanding of how populations may respond to climate change.

Recent data has shown that barnacle cyprid larvae, which are quick and easy to measure in the intertidal as representatives of marine larvae, settle in a synchronous and patchy manner along a rocky coast in Baja California (Ladah et al. 2005). Significant spatial variability in settlement and recruitment has been detected at scales down to < 100 m. A remaining question is whether this patchiness is driven by the nearshore physics of the region or by the behavior of the larvae, or both? Also, significant alongshore differences in nutrient concentrations between adjacent kelp beds (<1 km) have been detected (Ladah, unpublished data). For example, a simple physical mechanism that could explain this patchiness in settlement and in nutrients is the alongshore variable patterns of internal wave transport and mixing, presumably driven by local bathymetric differences, which could bring larvae and nutrients to the shore in a variable manner. Fluorescent dye, optically measured with a fluorometer, is commonly used as a tracer to study dispersion in marine (e.g., Okubo 1971) environments. Dye release and transport experiments are a relatively straightforward way to measure these physical transport mechanisms and can determine if biological and chemical patterns coincide with physical patterns of dye transport.

An important final topic to note is that the ability of internal waves to propagate into shallow coastal zones depends on multiple factors, central among which are the degree of water column density stratification and the topography of the coastal shelf (Baines 1997). As the ocean warms and wind patterns change, predictions of how coastal stratification will change are becoming available. In light of the current and predicted trend of a warming ocean (Fields et al. 1993), deepening thermocline depths, and reduced nutrient supply to the coast via coastal upwelling (Di Lorenzo et al. 2005, Auad et al. 2006), many organisms that depend on these transport mechanisms may be in danger if these mechanisms become reduced or disappear. Therefore, nearshore water column monitoring over the long term will become more and more valuable to understand the local and regional effects of a globally warming ocean.

Numerous other interesting questions will certainly arise from the high frequency monitoring of coastal zone biology and physics and how they interact. This remains a fairly understudied topic, although it is receiving much more attention in recent years, as physical oceanographers recognize the greater benefit of concurrent biological data to understand coastal integrative oceanography better. As biological oceanographers begin to approach physically inclined scientists, more and more interdisciplinary questions can be addressed. As biologists and physicists learn to communicate in the same language, a greater understanding of the coastal ocean will become evident. Essentially, important questions on connectivity between populations are becoming more and more feasible to address through this interdisciplinary approach to questions.

Macrofaunal Benthic Communities

Background

Estimates of the benthic macrofaunal community are often used to indicate environmental health because benthic animals a) are relatively sedentary (cannot avoid deteriorating water/sediment quality), b) have relatively long life spans (indicate and integrate water/sediment quality), c) consist of different species that exhibit different tolerances to stress (can be classified into functional groups), d) have an important role in cycling nutrients and other chemicals between the sediments and the water column and e) are commercially important or are important food sources for economically important species (Aller, 1978, 1982; Gray et al. 1988; Warwick et al. 1990; Dauer, 1993). Important progress has been achieved in the study of marine environment modifications by the use of different biological methods on benthic macrofauna. One of these methods is the use of species, for example macrobenthic invertebrates, as indicators of pollution. This method has been used in different regions (Leppaekoski, 1975; Rainbow, 1995, Boström et al., 2002, Diaz et al., 2004). Polychaetes are useful as a tool for environmental quality assessment (Pocklington and Wells, 1992).

In aquatic ecosystems, ecological processes such as organic matter mineralization and nutrient cycling are regulated by microbial activities in sediments (Smith, 1989; Sundbäck et al., 2004). Benthic animals interact with microbes and modify the biogeochemistry of sediments (Palmer et al., 1997). Many examples have highlighted the key ecological role played by invertebrate bioturbation in marine environments (e.g. Aller, 1988; Krantzberg, 1985; Posey, 1990; Pearson, 2001). By redistributing particles and modifying water fluxes at the water-sediment interface, annelids modify the physico-chemical and biological properties of the substratum and the interstitial water (Aller, 1994; Rosenberg, 2001; Gilbert et al., 2003; Stief et al., 2004).

Polychaetes play a major role in the functioning of benthic communities, in terms of recycling benthic sediments and in the burial of organic matter. Dense tube mats may promote the deposition and retention of high BOD organic matter (Rhoads and Germano, 1986). From a management perspective, polychaetes are useful organisms for identifying problem sites and for the assessment of the severity of the problem. They respond to disturbance induced by different kinds of pollution, by exhibiting quantitative changes in assemblage distribution. Polychaetes can also be used as indicators of recovery of benthic environments from perturbations since in many cases there are major elements of the recolonization process and of major importance to the initial structuring of the community (Díaz-Castañeda and Reish, in press).

Some species of the families Capitellidae, in particular Capitella complex, Cirratulidae and Spionidae have become widely accepted as organic pollution indicators (Tsutsumi 1990, Pocklington and Wells, 1992; Ajmal-Khan et al., 2004). The absence of genus Lumbrineris has also been documented as indication of pollution (Ryggs, 1985).

General Methods

The knowledge of the composition and structure of benthic communities give us insight into the biological diversity of sea-floor communities and the ecosystem energy flow. Benthic fauna provide fundamental data that are relevant to general objectives of most marine monitoring programs. In addition, they are especially important components of monitoring because they are sedentary and they respond to pollutant stress (Bilyard, 1987).

Marine sediments were collected with a van Veen grab (0.1 m^2) in Bahía Todos Santos and Salsipuedes; with a Smith-McIntyre grab (0.1 m^2) in Bahia Magdalena and a box corer (0.04 m^2) in San Quintin. Sediments were sieved using a 1.0 mm mesh size and retained material was fixed in a 7% buffered sea-water–formalin solution. In the laboratory, samples were washed using a 0.5 mm sieve and transferred to 70% ethanol. Macrofaunal organisms were sorted and counted; polychaetes were identified to the species level. Organic matter content was determined by ignition loss (Dean, 1974; Byers et al., 1978). At Salsipuedes, organic carbon and nitrogen were evaluated using the method of Hedges and Stern (1984) with an Elemental Analyzer LECO CHNS-932. Granulometry was determined by means of a Laser Horiba LA 910. The redox potential (Eh) was measured immediately after collection of each sample by probing 2-3 cm inside the sediments using an Ingold electrode coupled to a field potentiometer.

Olmstead and Tukey's test (Sokal and Rohlf, 1995) was applied to analyze spatial distribution of polychaetes species. This technique plots the frequency of appearance in each site sampled expressed as percentage against the density of organisms for each species. A mean average was calculated for both axes, resulting in four quadrants: I—Frequent and abundant species, II—Non-frequent and abundant species, III—Non-frequent and non-abundant species, and IV—Frequent and non-abundant species. Statistical methods were used to describe the structure and organization of the polychaetes communities within the bays. Diversity was calculated using Shannon's Index (Shannon and Weaver, 1963; Frontier, 1985) and Pielou's evenness index (Pielou, 1977). Stress predictability modeling (Alcolado 1992) was applied to establish the level of environmental stress existing in these bays. Environmental severity or stress was predicted based on values of diversity (H') and evenness (J'), coupled with redox potential values.

Ordination and classification methods were used to detect spatial patterns among the polychaete fauna. Cluster analysis using Sorensen, Jaccard and Bray-Curtis similarity coefficients (Bray and Curtis 1957; Sokal and Rohlf 1995) were employed to evaluate the level of association of different sampling sites (stations). The raw data were transformed by using log (x+1) or square root as suggested by Frontier (1983) and Legendre and Legendre (1984). Non metric Multidimensional Scaling (MDS) was used with the Program PRIMER 5.1, Plymouth Marine Laboratory, since this technique has demonstrated to be suitable for multiple ecological purposes (Clarke 1993). It is based on an association matrix containing the similarities between all pairs of samples; when samples are close to each other, they have more similar faunistic profiles. The quality of this representation is measured by the "stress"-value. The lower this value the better the representation.

Results

Data collected from soft-bottom environments of four sites along Baja California are presented; the case studies correspond to four bays located along the western coast of Baja California peninsula: Bahía Todos Santos, Bahía San Quintín, Bahía Magdalena and Bahía Salsipuedes.

Bahía Todos Santos

Bahía Todos Santos is located about 100km south of the United States—Mexico border, in the western coast of Baja California, between Punta Banda and Punta San Miguel at 31° 40'–31° 56' N and 116° 36'–116° 50' W. We analyzed macrofauna diversity patterns in Todos Santos Bay, Baja California, Mexico (Díaz-Castañeda and Harris, 2004). In 1994 thirty-nine stations were sampled (Figure 7a).

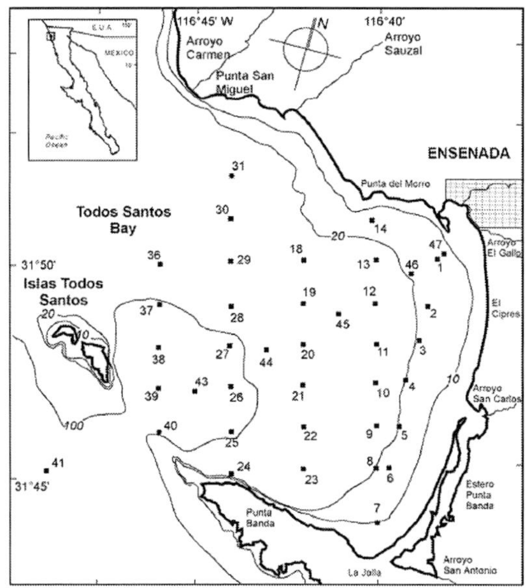

(a)

Figure 7. (Continues)

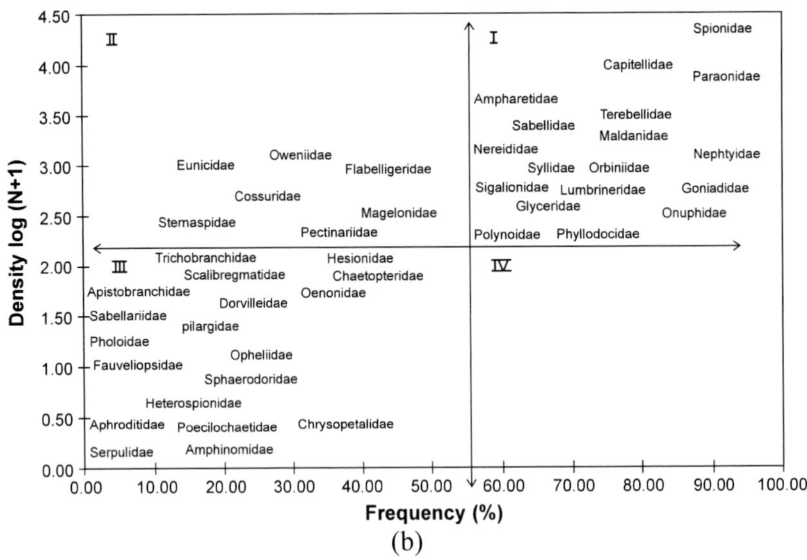

(b)

Figure 7. Study area in Todos Santos Bay with sampling stations (a). Olmstead and Tukey's analysis placed polychaetes families into three quadrants corresponding to dominant, restricted and rare (b).

Measurements of depth, temperature, salinity, dissolved oxygen; organic content and sediment particle size were made. Depth ranged between 7 - 400m and percentage of organic matter in sediments between 0.6-10.6%. Organisms belonging to 13 Phyla were collected: Annelida, Echinodermata, Phoronida, Brachiopoda, Mollusca, Bryozoa, Arthropoda (Crustacea), Nemertea, Sipunculida, Cnidaria, Porifera, Phoronida and Nematoda. Most abundant groups were Polychaeta (13,757 organisms), Amphipoda (2,930), Tanaidacea (1,581), Ostracoda (1,046), Echinodermata (926), Nemertea (828), Nematoda (720), Gastropoda (400), Bivalvia (342), Scaphopoda (96), Sipunculida (132).

Polychaetes constituted 64% of all invertebrate macrofauna, with 13,757 specimens in 44 families representing 203 species (Table 1). Three families are reported as new records for Baja California: Aphroditidae (*Aphrodita cf japonica*), Pholoidae (*Pholoe glabra, Pholoe* sp.), and Fauveliopsidae (*Fauveliopsis gabra).*

The best represented families were Spionidae, Capitellidae, Paraonidae, Cirratulidae, Maldanidae, Ampharetidae and Nephtyidae (Figure 7b). Todos Santos presented high species richness (species/station); values varied between 6 (near the harbor) and 67 species (next to Estero Punta Banda). Higher species richness values (48 to 67 sp/station) were located in the southern section of the bay. Abundances (individuals/station) were generally high (120–1434) except for some coastal stations.

Nearly one third of the stations presented Shannon index values higher than 4.00. Diversity (H') values ranged from 2.06 to 4.80; higher diversity values were found in the southern section of the bay (Figure 8). In most of the stations, Pielou's evenness index (J") showed high values: 58% of the stations presented values higher than 0.80. However, some coastal stations (2, 4, 7, 9, and 14) located close to the harbor and creeks seem relatively affected by freshwater discharges and pollution; J' values ranged between 0.54 and 0.67. These stations were characterized by lower abundance and diversity, and an increase in species belonging to family Spionidae.

Table 1. Polychaete species from Todos Santos Bay, Baja California (Diaz-Castañeda and Harris, 2004)

POLYCHAETES

AMPHARETIDAE
Amage anops
Amage sp. UI
Ampharete acutifrons
Ampharete finmarchica
Ampharete labrops
Ampharetidae sp. SD 1
Ampharetidae, UI
Amphicteis scaphobranchiata
Amphicteis sp.
Anobothrus bimaculatus
Asabellides lineata
Eclysippe trilobata
Lysippe sp. A
Lysippe sp. B
Melinna heterodonta
Melinna oculata
Mooresamytha bioculata
Paramage scutata
Sabellides manriquei
Schistocomus hiltoni
Schistocomus sp. A
Sosane occidentalis

APISTOBRANCHIDAE
Apistobranchus sp.

CAPITELLIDAE
Anotomastus gordiodes
Capitella complex sp.
Decamastus gracilis
Heteromastus sp.
Mediomastus sp.
Mediomastus acuta
Mediomastus ambisetus
Mediomastus californiensis
Notomastus latericeus
Notomastus sp. A
Notomastus sp., UI
Lumbrineris californiensis
Lumbrineris index
Lumbrineris limicola
Ninoe tridentata
Scoletoma tetraura
Lumbrineridae, UI

MAGELONIDAE
Mageiona hartmanae
Mageiona sacculata
Mageiona sp. B

MALDANIDAE
Axiothella TS sp. 1
Clymenella sp. A
Clymenella sp.
Clymenura gracilis
Euclymeninae sp. A
Euclymeninae sp. B
Euclymeninae UI
Maldane sarsi
Metasychis disparidentata
Notoproctus sp. A
Petaloclymene pacifica
Petaloproctus sp. A
Petaloproctus sp. B
Praxillella pacifica
Rhodine bitorquata

NEPHTYIDAE
Aglaophamus dicirrus
Aglaophamus erectans
Aglaophamus eugeniae
Aglaophamus sp.
Nephtys caecoides
Nephtys californiensis
Nephtys cornuta
Nephtys ferruginea
Nephtys sp.

OENONIDAE
Arabella TS sp. 1
Aricidea wassi
Aricidea sp. SD sp. 1
Cirrophorus branchiatus
Cirrophorus furcatus
Levinsenia gracilis
Paradoneis eliasoni
Paraonidae, UI

PECTINARIIDAE
Pectinaria californiensis

PILARGIDAE
Ancistrosyllis cf. groenlandica
Sigambra tentaculata
Sigambra sp

POECILOCHAETIDAE
Poecilochaetus sp.

SABELLIDAE
Jasmineira sp. B
Pseudofabricola californica

SABELLARIIDAE
Neosabellaria cementarium
Sabellaria alcocki-complex

SPIONIDAE
Apoprionospio pygmaea
Apoprionospio sp
Dipolydora cardalia
Dipolydora socialis
Laonice cirrata
Laonice nuchala
Laonice sp
Malacoceros punctata
Microspio pigmentata
Paraprionospio pinnata
Paraprionospio sp
Polydora cirrosa
Polydora sp.
Prionospio dubia

CALAMYZIDAE
Calamyzas amphictenicola

CHAETOPTERIDAE
Phyllochaetopterus limicolus
Spiochaetopterus sp. A

COSSURIDAE
Cossura modica
Cossura rostrata
Cossura sp. A
Cossura sp., UI

FLABELLIGERIDAE
Flabelligeridae, UI
Pherusa neopapillata
Piromis sp. A

GLYCERIDAE
Glycera americana
Glycera lapidum-complex
Glycera macrobranchia
Glycera nana
Glycera oxycephala
Glycera sp. B
Glycera sp. C
Glycera sp., UI

GONIADIDAE
Glycinde armigera
Goniada brunnea
Goniada littorea
Goniada maculata

HESIONIDAE
Gyptis sp. A
Heteropodarke heteromorpha
Micropodarke sp. A
Podarkeopsis sp. A
Podarkeopsis sp. D

LUMBRINERIDAE
Lumbrinerides platypygos
Drilonereis sp. A
Notocirrus californiensis

ONUPHIDAE
Diopatra ornata
Diopatra splendidissima
Diopatra tridentata
Diopatra sp.
Hyalinoecia juvenalis
Mooreonuphis nebulosa
Mooreonuphis veleronis
Mooreonuphis TS sp. 1
Mooreonuphis TS sp. 2
Nothria occidentalis
Onuphis elegans
Onuphis iridescens
Onuphis TS sp. 1
Onuphidae, UI
Paradiopatra parva
Paradiopatra TS sp. 1
Rhamphobranchium longisetosum

ORBINIIDAE
Leitoscoloplos pugettensis
Orbiniidae, UI
Phylo felix
Phylo sp.
Scoloplos acmeceps
Scoloplos TS sp. 1

OWENIIDAE
Myriochele gracilis
Myriochele sp. B
Myriochele sp. M
Owenia collaris

PARAONIDAE
Acmira catherinae
Acmira horikoshii
Acmira lopezi
Acmira simplex
Allia antennata
Allia sp. A SCAMIT
Allia TS sp. 1
Prionospio jubata
Prionospio lighti
Prionospio multibranchiata
Prionospio sp
Spio maculata
Spiophanes berkeleyorum
Spiophanes bombyx
Spiophanes duplex
Spiophanes duplex
Spiophanes fimbriata
Spiophanes sp.

STERNASPIDAE
Sternaspis fossor

SYLLIDAE
Autolytus sp.
Ehlersia heterochaeta
Exogone breviseta
Exogone dwisula
Exogone lourei
Eusyllis habei
Eusyllis transecta
Eusyllis sp.
Odontosyllis phosphorea
Pionosyllis articulata
Pionosyllis sp.
Sphaerosyllis ranunculus
Syllides mikeli
Syllidae UI
Typosyllis sp. TS 1
Typosyllis sp. TS 2

TEREBELLIDAE
Amaeana occidentalis
Eupolymnia heterobranchia
Lanassa venusta
Lanice conchilega
Phisidia sp. A
Pista bansei
Pista brevibranchiata
Pista moorei
Pista sp. C
Polycirrus sp. A
Polycirrus sp. TS 1
Polycirrus sp., UI

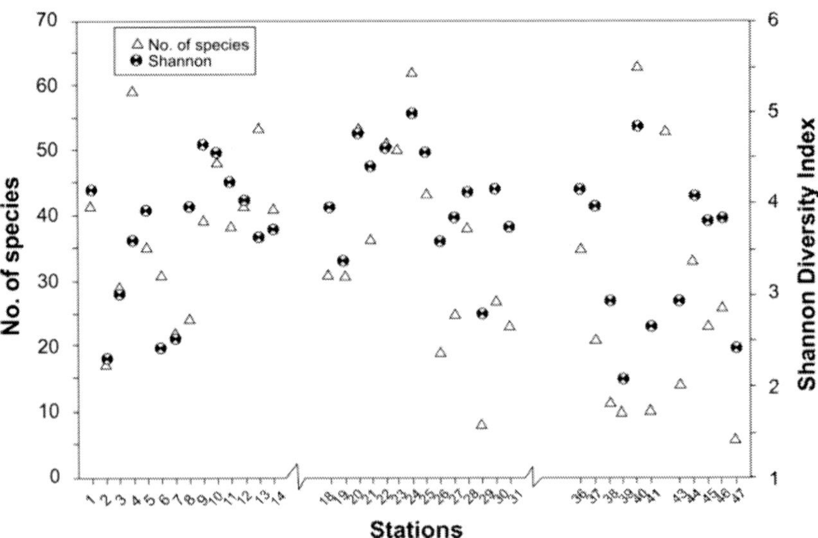

Figure 8. Species richness and diversity (H') at Todos Santos Bay, Baja California.

The dominant species in this bay belong mostly to the families Spionidae, Capitellidae, Paraonidae, Cirratulidae, Maldanidae, Orbiniidae, Nepthyidae, and Ampharetidae. Spionids were the most abundant polychaetes and included 16 taxa. Among the numerically dominant species were *Mediomastus* spp., *Apoprionospio pygmaea, Nepthys cornuta*, Lumbrinereidae UI, *Leitoscoloplos pugettensis, Cirratulus cirratus, Aricidea (Aricidea) wassi, Praxillella pacifica, Spiophanes bombyx, S. duplex, Prionospio* spp, *Exogone lourei, Pectinaria californiensis, Pherusa neopapillata, Amphicteis scaphobranchiata, Glycera armigera, A. (Acmira) catherinae, A. (Acmira) simplex, A.* (*Allia*) sp A, *Levinsenia gracilis*.

Analyzing the trophic composition, we found that 91 species were surface and subsurface deposit-feeders, 50 species carnivores, 36 species filter-feeders, and 24 species herbivores. By far the dominant trophic group corresponds to surface and subsurface deposit-feeders who exploit organic matter and its associated bacterial populations. Frequent upwelling allows an important primary productivity, part of which arrives to the ocean floor fueling benthic communities.

About 75% of Todos Santos Bay has high values of diversity, high values of evenness, and high species richness, all indicative of a favorable environment for polychaete development and a healthy benthic community (Snelgrove et al., 1997).

MDS analysis (see Díaz-Castañeda and Harris 2004) showed the separation between coastal-shallow stations, northwest stations characterized by the coarsest sediments, southern middle stations where the bivalve culture takes place and the submarine canyon (stations with more than 80% of silt-clay). Probably the biodeposition due to *Mytilus* cultures located in the southern section of the bay contribute partially to the high values of diversity and abundances found in that area. Todos Santos has a rich annelid fauna of which annelid polychaetes are the most important macrofauna group in terms of abundance and number of species, being extensively distributed in this bay. Their distribution seems strongly related to location in relation to the coast-line, sediment characteristics, and content of organic matter. Approximately 80% of stations were favorable environments for polychaetes, especially in the southern area where the bivalve culture takes place.

Bahía San Quintin

San Quintin is a 4,800 ha, shallow water embayment which has the shape of an inverted "Y" and consists of two basins: Bahía Falsa (west) and San Quintín (east). It is located about 300 km south from the United States-Mexico border. It is of volcanic origin and is situated on the northwestern coast of Baja California Peninsula, between 30°24'–30°30' N and 115°57'–116°01' W (Figure 9a). It is a temperate climate marine lagoon system relatively unimpacted, considered ecologically important because it is a nursery area for many species, a resting site for migrating birds which have lost most of their resting and feeding areas in North America, and its high productivity and diversity, in part due to periodic upwellings which supply nutrients (Alvarez-Borrego *et al*, 1975; 1976) and the abundance of marine macrophytic vegetation. The marine vegetation of San Quintin consists mainly of eelgrass *Zostera marina* best developed in the middle parts of the bay (Ibarra-Obando *et al*, 2004).

Bahía Falsa is the western basin; it communicates with the sea through a wide channel (approx. 420 m). Depth ranges between 1-15m. In this area oyster culture has taken place for more than two decades (Díaz-Castañeda and Rodríguez, 1998; Díaz-Castañeda *et al.*, 2005). As development pressure continues in Baja California coastal zone, there is an urgent need to have baseline data as well as understand ecosystem functioning in less disturbed conditions such as in San Quintin Bay.

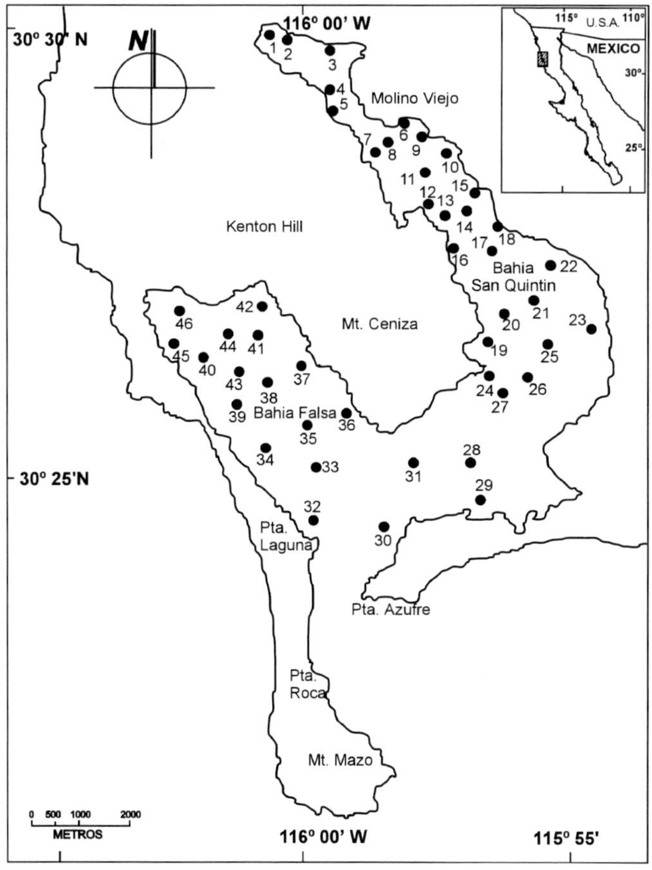

Figure 9 (Continues)

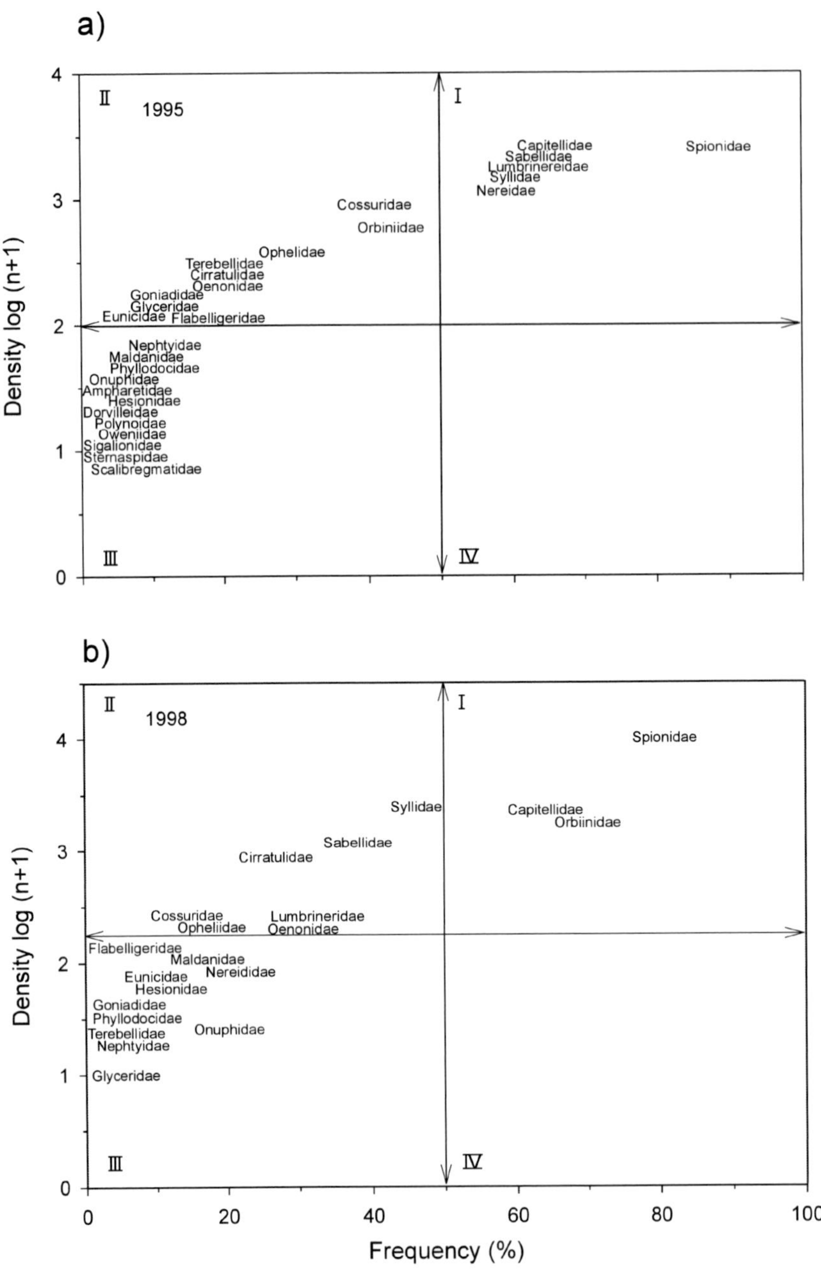

Figure 9. Sampling stations in San Quintín Bay. Frequency (%) and density (pol/0.1 m2) of polychaete families using Olmstead and Tukey's technique: a) 1995 b) 1998.

Macrofauna

Organisms belonging to nine Phyla have been collected: Annelida, Echinodermata, Mollusca, Bryozoa, Arthropoda (Crustacea), Nemertea, Sipunculida, Cnidaria, Nematoda. Most abundant groups were Polychaeta, Amphipoda, Tanaidacea, Ostracoda, Echinodermata, Gastropoda and Bivalvia.

Best represented polychaete families in 1995 were Spionidae, Capitellidae, Sabellidae, Lumbrineridae, Syllidae and Nereididae; while in 1998 we had Spionidae, Capitellidae, Orbiinidae and Syllidae (Figure 9b).

The diversity patterns of the polychaete fauna from this lagoon have been described (Díaz-Castañeda *et al.*, 2005); these benthic organisms can be used as environmental health indicators (Rice, 2003; Pocklington and Wells, 1992). Polychaetes were collected in 1995 and 1998. 46 stations were sampled (Figure 9b). A total of 3,275 polychaetes, 28 families, 56 genera, and 104 species were identified in 1995, and 3,168 polychaetes were collected in 1998, representing 21 families, 39 genera and 65 species (See Table 2). From all the macrofauna collected in both surveys, polychaetes represented 45.2%. From the species collected, 55% correspond to new records for the area. Families Dorvilleidae, Polynoidae, Oweniidae, Scalibregmatidae, Sternapsidae and Sigalionidae present in 1995 were not in 1998 survey. The stations with higher abundances (100 specimens/0.02 m2) were located on the southern half of Bahía San Quintin. Species richness and diversity were also higher in San Quintin arm (Díaz-Castañeda *et al.*, 2005). From the 30 families previously reported for San Quintin lagoon, 23 have been collected and 6 families were added: Ampharetidae, Oweniidae, Scalibregmatidae, Sternapsidae, Dorvilleidae and Sigalionidae. Families not found in both surveys were: Paraonidae, Magelonidae, Apistobranchidae, Sphaerodoridae, Trichobranchidae, Chrysopetalidae and Arenicolidae.

In 1995, the redox potential values (Eh) were negative in most of the stations. In the eastern arm (SQ) they varied between -340 and +162 mV; the western arm (BF) presented values between - 320 to +161 mV. Sediment temperatures oscillated between 19.1 and 22.0°C. Organic matter values varied between 0.3 to 3.4% in SQ and 0.1 to 3.1% in BF. In 1998 only half of stations were measured for Eh and temperature. In the eastern arm the Eh varied between -336 mV and +154, while the western arm presented values between - 308 and +187 mV. Sediment temperatures were in the range 19.8 to 22.1°C, while the organic matter content ranged between 0.5 to 4.0% in BSQ and 0.3 to 4.1% in BF. Results showed slightly lower redox potential values (-336 to +187 mV), slightly higher sediment temperatures (19.8°–22.1°C) and organic matter contents (0.3–4.1%) in 1998.

From 1995 to 1998 a change in the composition and structure of the polychaetes communities was noted; species richness decreased from 104 to 65 species. Bray-Curtis dendrogram using data from 1995 showed a clear separation between both arms (Díaz-Castañeda *et al.*, 2005). The trophic complexity changed between 1995 and 1998 with an increase of deposit-feeders, the abundance of other trophic categories decreased, indicating a loss of complexity. Significant changes in the abundance of some families were detected, some increased their abundances: Spionidae from 17% to 48%, Orbiniidae from 4% to 13%; other families decreased in terms of abundance and number of species: Lumbrinereidae from 11% to 1.4%, Nereididae from 9% to 1% and Sabellidae from 14% to 5%. These modifications altered the composition and structure of polychaete communities. Increased anthropogenic disturbance (oyster culture, agriculture) and environmental variability due to the ENSO 97–98 may have affected recruitment and survival of some macrobenthic species.

Table 2. Polychaete species collected at San Quintin Bay, Baja California (Diaz-Castañeda et al. 2005)

AMPHARETIDAE
Ampharete labrops Hartman, 1961
Ampharete sp Malmgren 1866
Amphicteis acutifrons Grube, 1850
Amphicteis sp Grube, 1850

CAPITELLIDAE
Capitella capitata Fabricius, 1780
Mediomastus californiensis Hartman, 1944
Mediomastus sp Hartman, 1944
Notomastus magnus Hartman, 1947
Notomastus tenuis Moore, 1909
Notomastus sp Sars, 1851

CIRRATULIDAE
Aphelochaeta marioni Saint-Joseph, 1894
Aphelochaeta Blake, 1991
Cirriformia spirabranchia Moore, 1904
Monticellina tesselata Hartman, 1960
Protocirrineris socialis Blake, 1996
Protocirrineris sp

COSSURIDAE
Cossura candida Hartman, 1955
Cossura sp A Webster & Benedict, 1887

DORVILLEIDAE
Dorvillea sp Parfitt, 1866

EUNICIDAE
Lysidice ninetta Verril, 1900
Marphysa disjuncta Hartman, 1961
M. sanguinea Montagu, 1815
Marphysa sp Quatrefages, 1865

FLABELLIGERIDAE
Pherusa capulata Moore, 1909
Piromis arenosus Kinberg, 1867
Piromis sp

GLYCERIDAE
Glycera americana Leidy, 1855
G. tenuis Hartman, 1944

GONIADIDAE
Goniada brunnea Treadwell, 1906
Goniada littorea Hartman, 1950

HESIONIDAE
Podarkeopsis glabra Hartman, 1961
Podarke pugettensis Johnson, 1901

LUMBRINERIDAE
Scoloioma crassidentata Fauchald, 1970
S. erecta Moore, 1904
S. monroi Fauchald, 1970
S. tetraura Schmarda, 1860

MALDANIDAE
Axiohtella rubrocincta Jonson, 1901
Axiothella sp Verril, 1900
Clymenura gracilis Moore, 1923
Euclymeninae sp A Ardwidsson, 1906
Isocirrus longiceps Moore, 1923
Maldane sp Grube, 1860

NEPHTYIDAE
Nephtys caecoides Hartman, 1938
Nephtys sp

NEREIDIDAE
Neanthes caudata delle Chiaje, 1828
Nereis latescens Chamberlin, 1919
N. pelagica Linné, 1758
Nereis sp Linné, 1758
Platynereis bicanaliculata Baird, *1863*
P. marphysa

OENONIDAE
Arabella iricolor Montagu, 1804
A. pectinata Fauchald, 1970
Drilonereis falcata Moore, 1911
D. longa Webster, 1879
D. mexicana Fauchald, 1970
Drilonereis sp. Claparède, 1870
Notocirrus californiensis Hartman, 1944

ONUPHIDAE
Kinbergonuphis sp Fauchald, 1982

OPHELIDAE
Armandia bioculata Hartman, 1938
A. brevis Moore, 1906
Ophelia pulchela Tebble, 1953
Polyophthalmus pictus Dujardin, 1839

ORBINIDAE
Leitoscoloplos mexicanus
L. normalis Day,1977
Naineris grubei Gravier, 1908
Phylo felix Kinberg, 1866
P. ornatus Verril, 1873
Scoloplos acmeceps Chamberlain, 1919
S. armiger Müller, 1776
S. ohlini Ehlers, 1901
S. texana Maciolek & Holland, 1978

OWENIIDAE
Owenia collaris Hartman, 1955

HYLLODOCIDAE
Eteone pacifica Hartman, 1936
Eteohe sp Savigny, 1820
Eulalia bilineata. Johnston, 1840
Eumida sp Malmgren, 1865

POLYNOIDAE
Harmotoe imbricata Linné, 1767
Harmotoe sp. Kinberg, 1855

SABELLIDAE
Chone infundibuliformis Kröyer, 1856
C. mollis Bush, 1904
Fabricinuda limnicola Hartman, 1951
Megalomma pigmentum Reish, 1963

SCALIBREGMATIDAE
Scalibregma sp Rathke, 1843

SIGALIONIDAE
Sthenelais fusca Johnson, 1897

SPIONIDAE
Aporionospio pigmaeus Hartman, 1961
Boccardiella hamata Webster, 1879
Microspio pigmentata Reish, 1959
Minuspio cirrifera Wirén, 1883
Polydora socialis Schmarda, 1861
P. websteri Hartman, 1943
Prionospio heterobranchia Reish, 1959
P. lighti Maciolek, 1985
Pseudopolydora paucibranchiata Okuda, 1937
Scolelepis squamata O. M. Müller, 1806
Spiophanes bombyx Claparède, 1870
S. duplex Chamberlain, 1919
S. missionensis Hartman, 1941
Spio pacifica
Spio sp

SYLLYDAE
Cicese sphaeroyilliformis Diaz & San Martin, 2001
Eusyllis sp. Malmgren, 1867
Exogone lourei Berkeley & Berkeley, 1938
Grubeosyllis mediodentata Westheide, 1974
Pionosyllis sp. Malmgren, 1867
Sphaerosyllis californiensis Hartman, 1966
Syllis aciculata Treadwell, 1945
S. gracilis Grube, 1840
S. heterochaeta Moore, 1909
Syllis sp Savigny, 1818

TEREBELLIDAE
Eupolymnia nebulosa Montagu, 1818
Pista alata Moore, 1909
Pista sp. Malmgren, 1865
Polycirrus sp. Grube, 1860

Table 3. Polychaete species found at Magdalena Bay, Baja California Sur (Diaz Castañeda and De Leon González, 2007)

Amphinomidae
Pareurythoe californica
Chloeia pinnata

Capitellidae
Heteromastus filiformis
Anotomastus gordiodes
Mediomastus sp
Mediomastus californiensis
Notomastus magnus
Notomastus tenuis
Notomastus sp

Chaetopteridae
Chaetopterus variopedatus
Spiochaetopterus costarum

Cirratulidae
Aphelochaeta glandaria
Aphelochaeta monilaris
Caulleriella alata
Caulleriella cristata
Caulleriella hamata
Caulleriella pacifica
Chaetozone setosa
Chaetozone corona
Chaetozone gracilis
Chaetozone hartmanae
Chaetozone spinosa
Monticellina cryptica
Monticellina siblina
Monticellina serratiseta
Monticellina tesselata
Cirratulus cirratus

Dorvilleidae
Dorvillea annulata
Protodorvillea kefersteini

Glyceridae
Glycera americana
Glycera tesselata

Goniadidae
Glycinde armigera
Goniada maculata

Lumbrineridae
Lumbrinerides platypygos
Lumbrineris cruzensis
Lumbrineris latreilli
Scoletoma sp

Magelonidae
Magelona californica
Magelona sp

Nephtyidae
Aglaophamus verrilli
Aglaophamus sp
Nephtys cornuta
Nephtys simoni

Nereididae
Nereis procera
Neanthes succinea

Oenonidae
Arabella iricolor

Onuphidae
Mooreonuphis nebulosa
Onuphis iridescens
Diopatra tridentata

Opheliidae
Armandia brevis
Armandia cf agilis

Orbiniidae
Leitoscoloplos kerguelensis
Scoloplos texana
Scoloplos armiger
Phylo felix
Scoloplos (Leodamas) sp

Oweniidae
Owenia collaris

Paraonidae
Aricidea (Acmira) catherinae
Aricidea (Acmira) lopezi
Aricidea (Allia) sp A
Aricidea (Aricidea) sp B

Polynoidae
Halosydna brevisetosa
Harmothoe lunulata
Malmgrenia sp
Cirrophorus aciculatus
Cirrophorus branchiatus

Maldanidae
Euclymene papillata
Praxillella gracilis

Phyllodocidae
Phyllodoce hartmanae
Phyllodoce medipapillata
Eumida sp
Eteone californica

Pilargiidae
Parandalia ocularis

Poecilochaetidae
Poecilochaetus johnsoni

Sabellidae
Megalomma pigmentum

Sigalionidae
Sigalion spinosus
Sthenelais tertiaglabra

Spionidae
Apoprionospio pygmaea
Dispio uncinata
Dipolydora socialis
Microspio sp.
Pseudolydora kempi
Paraprionospio pinnata
Polydora socialis
Prionospio (Minuspio) lighti
Prionospio (Minuspio) cirrifera
Prionospio (Prionospio) jubata
Prionospio ehlersi
Prionospio steenstrupi
Scolelepis texana
Scolelepis squamata
Spiophanes missionensis
Spiophanes bombyx
Spiophanes duplex
Spiophanes fimbriata

Syllidae
Exogone lourei
Gyptis brunnea
Syllis (Typosyllis) sp

Serpulidae
Hydroides brachyacantha

Terebellidae
Polycirrus sp
Streblosoma sp B

Bahía Magdalena

Magdalena Bay is one of the most important bays in the western Mexican Pacific, because of its considerable area (650 km²), habitat diversity and high productivity which make it an essential region for important fisheries like sardines and tuna (Alvarez-Borrego *et al.*, 1975). It is a eutrophic subtropical bay located between 24° 15' - 25° 20'N and 111° 30' - 112° 15' W (Figure 10a) in the southern limit of the California Current; it is connected to the open sea through a tidal pass 30-40 m deep and 4 km wide (Zaytsev *et al.*, 2003).

It is among the largest remaining pristine coastal lagoon systems along the Pacific coast of the Americas and it is located in a transition zone between tropical and temperate environments. It is a natural reserve of grey whales that arrive each year, an important feeding and nursery area for various species of sea turtles and a sanctuary for resident and migratory birds. Inputs of nutrients from subsurface waters off the continental shelf are transported through the main entrance by tidal currents boosting phytoplankton production (Zaytsev *et al.*, 2003; Robinson *et al.*, 2007) and benthic communities (Díaz-Castañeda and de León-González, 2007)

Macrofauna

To study the composition and structure of benthic macrofauna, two surveys were conducted in 1996 and 1999 with El Puma (UNAM) and F. de Ulloa (CICESE) respectively, with 21 stations distributed mainly in the central area and the mouth of Magdalena Bay, Baja California Sur (Figure 10a). Polychaetes were one of the most abundant groups. Organisms from 10 Phyla were found: Annelida, Echinodermata, Mollusca, Arthropoda, Nemertea, Sipunculida, Bryozoa, Cnidaria, Nematoda and Chordata. Most abundant groups were Polychaeta, Amphipoda, Tanaidacea, Ostracoda, Echinodermata, Gastropoda and Bivalvia.

In 1996, 1640 polychaetes were collected; they represented 27 families, 53 genera and 87 species. The best represented families (more abundant) were Spionidae, Cirratulidae, Paraonidae, Opheliidae, Lumbrineridae, Orbiinidae and Nereididae (Figure 10b). More specious families were Cirratulidae, (13 species), Spionidae (11 species) and Paraonidae (9 species) (Díaz-Castañeda and de León-González, 2007). In 1999, 961 polychaetes were collected, belonging to 25 families, 38 genera and 59 species. Most specious families were Cirratulidae (13 species), Spionidae (11 species), Paraonidae (9 species) and Syllidae (5 species). In both surveys polychaetes represented 50% of all invertebrate macrofauna. The list of polychaetes species collected in both surveys is presented in Table 3.

Highest values of abundance were located in stations 1, 7, 8, 13 and 15 in the northwest and east of the bay's mouth. Higher species richness was located on the northwest area, reaching up to 18 families/station (station 13). Species richness varied between 8-33 species/station; highest species richness was found in the northwest section of the study area (stations 12, 13). In general, most specious families at Magdalena were surface and subsurface deposit-feeders, followed by carnivores and suspension-feeders. The stress-predictability modeling (Alcolado, 1992) characterized 30% of stations as presenting very favorable and stable conditions, 52% as constant with a degree of environmental stress and 18% as moderately favorable. The average value of organic matter in the bay was 2.3%.

Bray–Curtis coefficient separated 5 groups of stations in relation to their depth and location in the bay: mouth, near the mouth, northwestern area, central area and eastern area. Non metric multidimensional scaling analysis confirmed Bray-Curtis dendrogram, indicating a separation between the sampling sites of the mouth and near the mouth, from those on the eastern area and the western area of the bay.

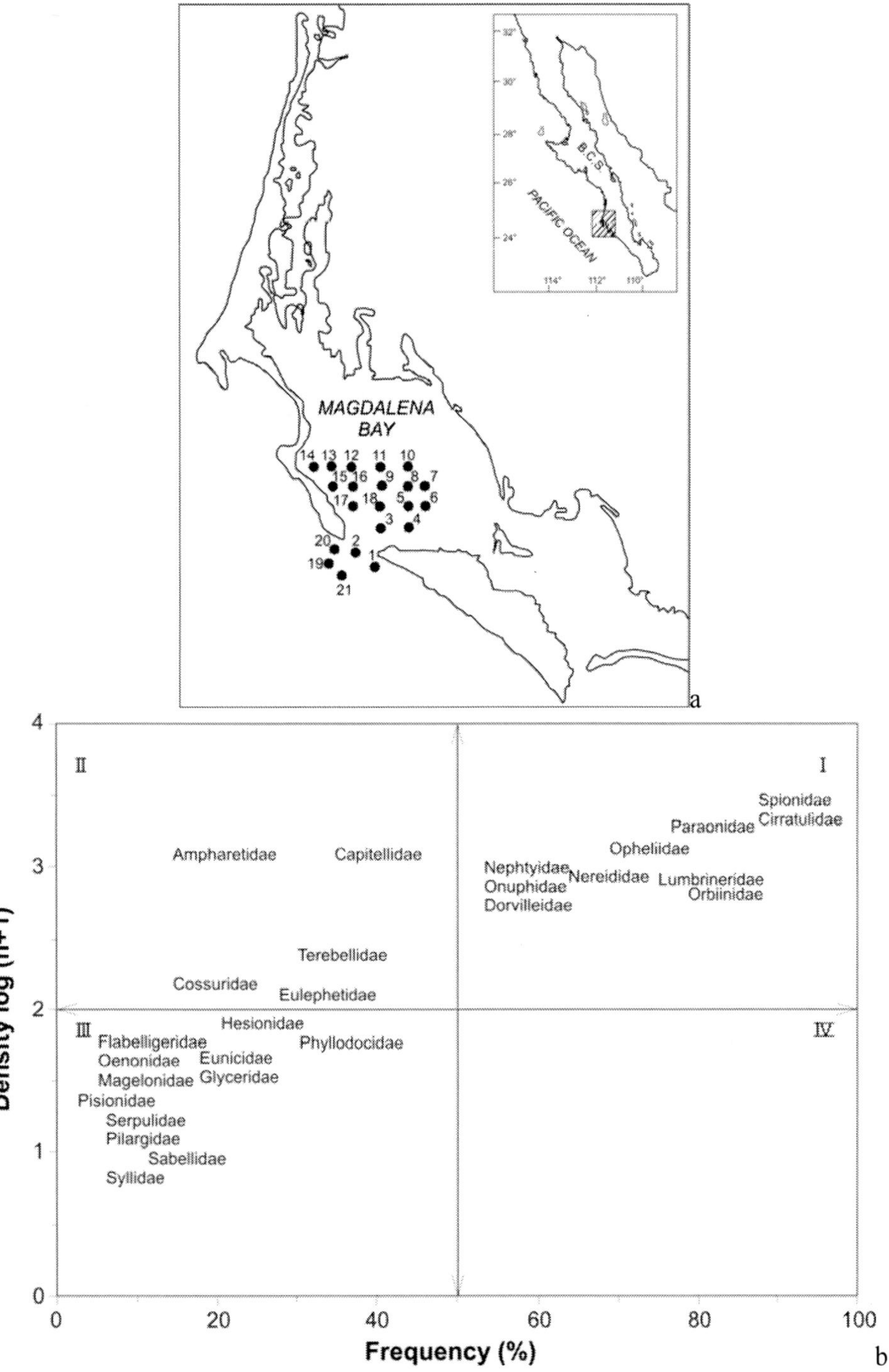

Figure 10. Study area at Magdalena Bay and location of sampling sites (a). Olmstead and Tukey's analysis placed polychaetes families collected in 1996 into three quadrants corresponding to dominant, restricted and rare (b).

MDS analysis (21 stations x 87 polychaete species) had a stress value of 0.11 (Figure 11) and showed 3 groups of stations: to the right the eastern section of Magdalena, to the left the western one and in the lower part 5 stations that correspond to the mouth of the bay. Station 19 located at only 23 m had the lowest percentage of organic matter (0.87%) and fine fraction (13.6%) isolated from the rest. Differences in Eh values, organic matter, granulometry, bathymetry and hydrodynamics favor different assemblages of polychaetes in different areas of this enormous bay. Highest values of abundance and diversity were found in the western part of Magdalena.

Upwelling and hydrological characteristics in this area favor the development of macrobenthos, polychaetes annelids predominate followed by peracarid crustaceans and bivalve molluscs. The Stress-predictability modeling showed that one third of stations correspond to Ambient 1 characterized as very favorable and stable, 52 % correspond to Ambient 3 characterized as constant with a degree of environmental stress and 18% correspond to Ambient 4 which is moderately favorable, with unstable conditions and a certain degree of environmental stress (Díaz-Castañeda and de León-González, 2007).

It is essential to sample the whole bay, particularly the shallower areas that could not be sampled in this study, in order to have complete information about the structure of benthic communities.

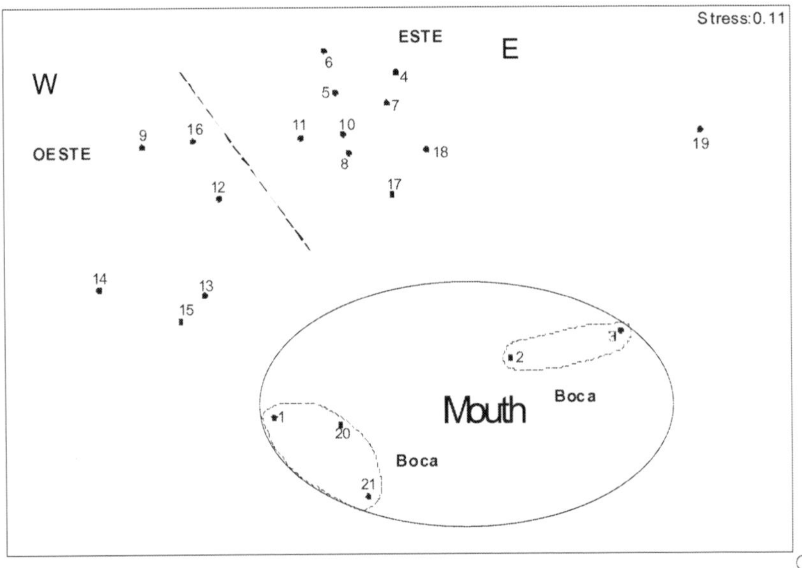

Figure 11. Non-metric multidimensional scaling (MDS) ordination of stations sampled at Magdalena, Baja California Sur.

Salsipuedes Bay

Salsipuedes Bay is a 30 km² bay located between 31° 92'-31° 98' N and 116° 85'-116° 75' W, at approximately 15 km north of Ensenada; tuna have been cultured there since 2002. In Baja California tuna fattening mariculture is expanding rapidly and there are increasing concerns about possible ecological impacts in the coastal zone.

Sediment samples were taken in 18 stations located in Salsipuedes Bay (Figure 12a).

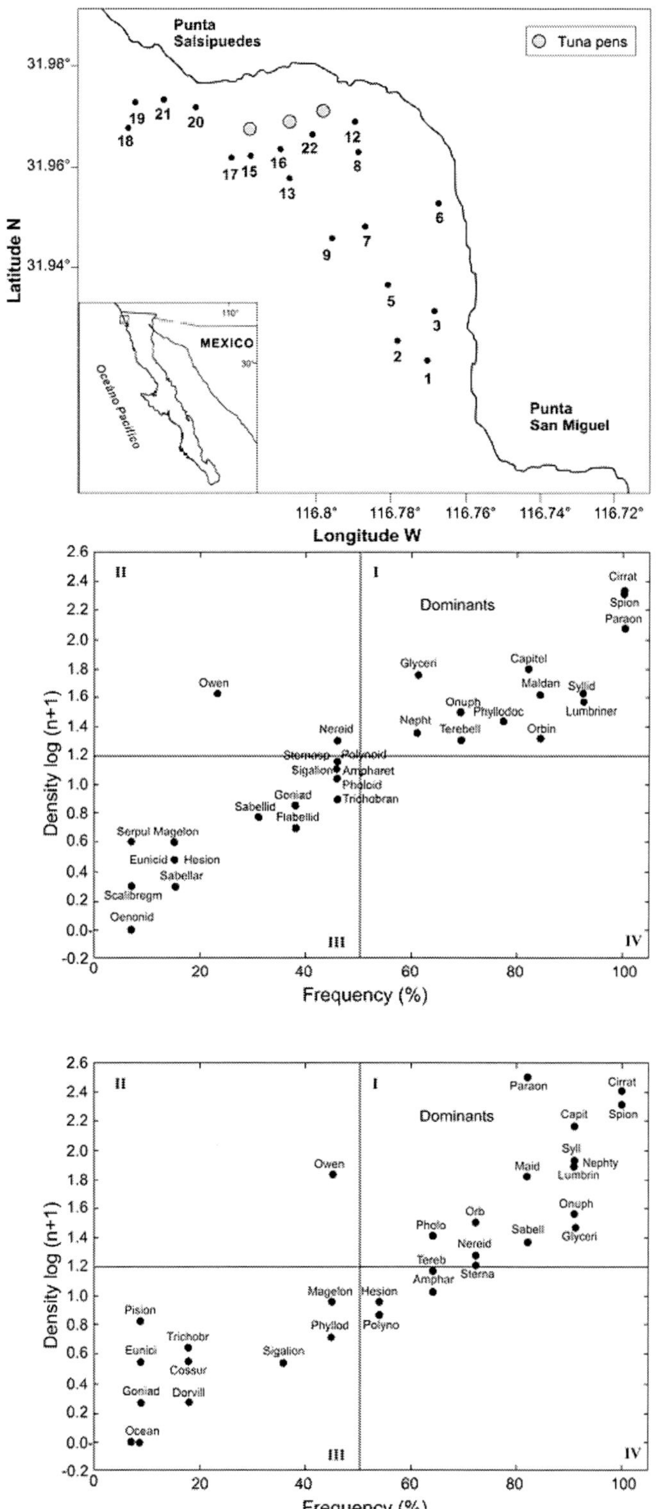

Figure 12. Location of sampling sites at Salsipuedes Bay, Baja California (a). Olmstead and Tukey's analysis placed polychaetes families (up 1995, down 1998) essentially into three quadrants corresponding to dominant, restricted and rare (b).

Table 4. Polychaete species collected at Salsipuedes Bay, Baja California

Paraonidae
Aedecira alisetosa Fauchald, 1972
Aricidea catherinae Laubier, 1967
Aricidea lopezi Berkeley & Berkeley, 1956
Aricidea simplex Bay, 1963
Aricidea antennata Annenkova, 1934
Aricidea ramosa Annenkova, 1934
Aricidea wassi Pettibone, 1965
Aricidea sp
Cirrophorus furcatus Hartman, 1957
Levinsenia gracilis Tauber, 1897
Levinsenia oculata Hartman, 1957
Paraonella sp
Spionidae
Apoprionospio pygmaea Hartman, 1961
Laonice nuchala Blake, new sp
Microspio pigmentata Reish, 1959
Microspio spinosa Blake, new sp
Minuspio lighti Maciolek, 1985
Paraprionospio pinnata Ehlers, 1901
Prionospio steenstrupi Malmgren, 1867
Polydora websteri Hartman, 1943
Spiophanes bombyx Claparede, 1870
Spiophanes duplex Chamberlain, 1919
Spiophanes kroeyeri Grube 1860
Cirratulidae
Aphelochaeta multifilis Moore, 1909
Aphelochaeta tigrina Blake, new sp
Monticellina cryptica Blake, new sp
Monticellina sp
Chaetozone corona Berkeley & Berkeley, 1941
Chaetozone hartmanae Hartman, 1963
Chaetozone senticosa Blake, new sp
Capitellidae
Mediomastus ambiseta Hartman, 1947
Mediomastus acuta Hartman, 1969
Mediomastus californiensis Hartman, 1944
Neonotomastus sp
Notomastus sp

Syllidae
Exogone breviseta Kudenov & Harris, 1995
Exogone lourei Berkeley & Berkeley, 1938
Brania californiensis Kudenov & Harris, 1995
Eusyllis habei Imajima, 1966
Pionosyllis articulata Kudenov & Harris, 1995
Pionosyllis sp
Sphaerosyllis californiensis Hartman, 1966
Sphaerosyllis sp
Ehlersia heterochaeta Moore, 1909
Nepthyidae
Aglaophamus eugeniae Fauchald, 1972
Nephtys californiensis Hartman, 1938
Aglaophamus dicirris Hartman, 1950
Aglaophamus verrili McIntosh, 1885
Nephtys caecoides Hartman, 1938
Nepthys sp
Ampharetidae
Amphareta acutifrons Grube, 1860
Amphicteis labrops Hartman, 1961
Amphicteis glabra Moore 1905
Asabellides lineata Berkeley & Berkeley, 1943
Melinna oculata Hartman, 1969
Sabellides cf manriquei Salazar-Vallejo, 1996
Sosane occidentalis Hartman, 1969
Cossuridae
Cossura candida Hartman, 1955
Cossura brunnea Fauchald, 1972
Cossura sp
Flabelligeridae
Pherusa neopapillata Hartman, 1961
Pherusa sp
Brada villosa, Rathke, 1843
Glyceridae
Glycera oxycephala Ehlers,1887
Glycera tenuis Hartman, 1944
Glycera tesselata Grube, 1863
Glycera americana Leidy, 1855
Glycera sp

Hesionidae
Podarkeopsis glabra Hartman, 1961
Podarkeopsis sp
Micropodarke sp
Lumbrineridae
Lumbrineris crasidentata Fauchald, 1970
Lumbrineris californiensis Hartman, 1944
Scoletoma tetraura
Ninoe tridentata Hilbig, 1995
Magelonidae
Magelona hartmanae Jones, 1978
Magelona pitelkai Hartman, 1944
Magelona sp
Maldanidae
Axiothella rubrocincta Johnson, 1901
Clymenura gracilis Hartman, 1969
Praxillela pacifica Berkeley, 1929
Praxillela sp
Petaloclymene pacifica Green, 1997
Notoproctus sp
Clymenella sp
Rhodine bitorquata Moore, 1923
Euclymeninae sp A SCAMIT, 1987
Maldane sarsi Malmgren, 1865
Oenonidae
Arabella iricolor Montagu, 1804
Notocirrus californiensis Hartman, 1944
Drilonereis mexicana Fauchald, 1970
Drilonereis sp
Onuphidae
Diopatra ornata Moore, 1911
Diopatra tridentata Hartman, 1944
Mooreonuphis nebulosa Moore, 1911
Mooreonuphis sp
Onuphis elegans Johnson, 1901
Onuphis iridescens Johnson, 1901
Onuphis sp
Hyalinoecia juvenalis Moore, 1911
Paradiopatra parva Moore, 1911

Ramphobrachium longisetosum Berkeley & Berkeley, 1938
Nothria occidentalis Fauchald, 1968
Oweniidae
Myriochele gracilis Hartman, 1955
Myriochele sp
Owenia collaris Hartman, 1955
Orbiniidae
Phylo felix Kinberg, 1866
Scoloplos acmeceps Chamberlin, 1919
Aricidea (Aedecira) pacifica Hartman, 1944
Aricidea (Allia) quadrilobata Webster & Benedict, 1887
Levinsenia gracilis Tauber, 1869
Pholoidae
Pholoe glabra Hartman, 1961
Sabellidae
Chone mollis Bush 1904
Megalomma pigmentum Reish, 1963
Sternapsidae
Sternapsis fossor
Terebellidae
Eupolymnia heterobranchia Johnson, 1901
Lanassa gracilis Moore, 1923
Lanassa venusta venusta Malm, 1874
Pista moorei Berkeley & Berkeley, 1942
Pista sp1
Nereididae
Neanthes arenaceodentata
Neanthes sf acuminata
Nereis sp
Polynoidae
Harmothoe multisetosa Moore 1902
Harmothoe imbricata Linnaeus 1767
Halosydna brevisetosa Kinberg 1855

Lepidonotus sp
Malmgreniella sp
Pisionidae
Pisione cf remota Southern 1914
Phyllodocidae
Eulalia californiensis Hartman 1936
Phyllodoce medipapillata Moore 1909
Phyllodoce sp
Trichobranchidae
Terebellides californica Williams 1984
Terebellides sp
Sigalionidae
Sigalion spinosus Hartman 1939
Sthenelais tertiaglabra Moore 1910
Eunicidae
Eunice cf multipectinata Moore 1911
Marphysa disjuncta Hartman 1961
Goniadidae
Goniada maculata Orsted, *1843*
Goniada littorea Hartman, *1950*
Glycinde sp
Dorvilleidae
Dorvillea (Schistomeringos) annulata Moore 1906
Dorvillea sp
Serpulidae
Hydroides pacificus Hartman 1969
"Pracosregus sp
Scalibregmatidae
Scalibregma californicum Blake 2000
Sabellariidae
Sabellaria gracilis Hartman 1944

Table 5. Collections visited to check records and coral species identification from the Mexican Pacific

Code	Name of collection	Museum/ Institution	Main institution
AHF	Collection of Marine Invertebrates. Seccion Cnidaria	Allan Hancock Foundation	University of Southern California
CASIZ	Invertebrate Zoology Collection	Invertebrate Department	California Academy of Sciences
CCC CUC	Colección Científica de Cnidarios	Centro Universitario de la Costa	Universidad de Guadalajara
CESEM-UAG	Colección de la Escuela Superior de Ecología Marina	Centro de Estudios Superiores	Universidad Autónoma de Guerrero
ECO-CH-C	Colección de Invertebrados	Museo de Historia Natural	El Colegio de la Frontera Sur
IC-LACM	Invertebrate Collection	Museum of Natural History	Los Angeles County Museum
IIO-UABC	IIO-UABC	Instituto de Investigaciones Oceanicas	Universidad Autónoma de Baja California
LACM	E.C. Wilson Collection	Museum of Natural History	Los Angeles County Museum
MHNUABCS	Museo de Historia Natural	Museo de Historia Natural	Universidad Autónoma de Baja California Sur
MHNUMAR-002	Colección Corales	Museo de Historia Natural	Universidad del Mar
MHSA	Museum History, Science and Art,	Museum of Natural History	Los Angeles County Museum
SIOCo	Benthic Invertebrate Collection. Coelenterata	SCRIPPS Institution of Oceanography	University of California, San Diego
UCLA	UCLA Collection	Museum of Natural History	Los Angeles County Museum
UCMP	Collection of Invertebrates	The Museum of Paleontology	University of California
USNM	Invertebrate Collection	National Museum of Natural History	Smithsonian Institution
Warner	Warner Collection	Museum of Natural History	Los Angeles County Museum
YPM	Collection of Invertebrates	Yale Peabody Museum	Yale University

Table 5. (Continued)

Code	City	State	Country
AHF	Los Angeles	California	USA
CASIZ	Los Angeles	California	USA
CCC CUC	Puerto Vallarta	Jalisco	México
CESEM-UAG	Chilpancingo	Guerrero	México
ECO-CH-C	Chetumal	Quintana Roo	México
IC-LACM	Los Angeles	California	USA
IIO-UABC	Ensenada	Baja California	México
LACM	Los Angeles	California	USA
MHNUABCS	La Paz	Baja California Sur	México
MHNUMAR-002	Puerto Angel	Oaxaca	México
MHSA	Los Angeles	California	USA
SIOCo	San Diego	California	USA
UCLA	Los Angeles	California	USA
UCMP	Berkeley	California	USA
USNM	Washington	DC	USA
Warner	Los Angeles	California	USA
YPM	New Haven	Connecticut	USA

Redox potential in 2003 ranged between -113 and -200 mV, while in 2004 it ranged between -110 and -302 mV. Organic carbon concentrations varied between 0.20-2.53 %, lowest values were located in the southern part of the bay (far from tuna pens); highest concentrations were found in the northern section, west of the tuna pens. Organic N varied between 0.02 to 0.12%, highest concentrations (0.07-0.12%) were located also in the northern section of the bay; stations situated at the south and near the coast presented the lowest N concentrations (0.02-0.04%).

A total of 9,291 organisms belonging to 7 Phyla were collected: Polychaeta, Mollusca, Crustacea, Echinodermata, Cnidaria, Sipuncula and Bryozoa. Polychaetes accounted for 62% of all invertebrate macrofauna, with 5,765 specimens representing 34 families and 149 species (Table 4). Polychaetes were dominant in almost all sampling stations. Best represented families in 2003 were Cirratulidae, Spionidae, Paraonidae, Capitellidae, Syllidae, Lumbrineridae and Maldanidae; in 2004 were Cirratulidae, Spionidae, Capitellidae, Nepthyidae, Glyceridae Paraonidae and Lumbrineridae (Figure 12b). Families with the highest species richness were Paraonidae (12 spp), Spionidae (11 spp.), Onuphidae (11 spp.), Maldanidae (10 spp.), Syllidae (9 spp), Cirratulidae and Ampharetidae with 7 spp. Among the most abundant species were *Aphelochaeta multifinis, Mediomastus ambiseta, Prionospio steenstrupi Spiophanes bombyx, Apoprionospio pygmaea, Paraonella sp, Monticellina sp, Aricidea (Allia) ramosa, Spiophanes bombyx, Spiophanes duplex and Levinsenia gracilis.*

109 and 94 species were collected in 2003 and 2004 respectively. This bay had a broad range of species richness per station. Values varied between 11-32 sp/station in 2003 and 7-29 sp/station in 2004. Higher species richness values were located mainly in the middle and southern sections of the bay. Abundance of polychaetes decreased in 2004. In 2003 Shannon index varied between 2.26-3.40 bits/ind. highest diversity values were located in the southern section of Salsipuedes bay. In 2004 diversity fluctuated between 2.31-3.35, highest values were found in the northern section (stations 8, 12, 17), south of the tuna pens. The stress-predictability modeling characterized 85% of stations in 2003 and 77% in 2004 as presenting favorable and stable conditions, the rest were considered moderately disturbed. Non-metric multidimensional scaling (MDS) revealed that the similarity of polychaetes community structure depended on the distance from tuna pens (Figure 13).

Our results show that some macrofaunal groups such as crustacean decapods, echinoderms, annelids and amphipods have increased their abundances, certainly exploiting the abundant organic matter which is accessible and come from tuna fattening facilities. Results also indicate that at Salsipuedes the dominant trophic group corresponds to deposit-feeders, followed by carnivores. Although we generally found negative redox potential values, this area is still a relatively favorable environment for macrofauna and particularly polychaetes; apparently local circulation has at least partially dispersed the excess organic matter. Nevertheless in some samples from 2006 we found the pollution indicator polychaete *Capitella* spp and the increase of abundance of three *Mediomastus* species.

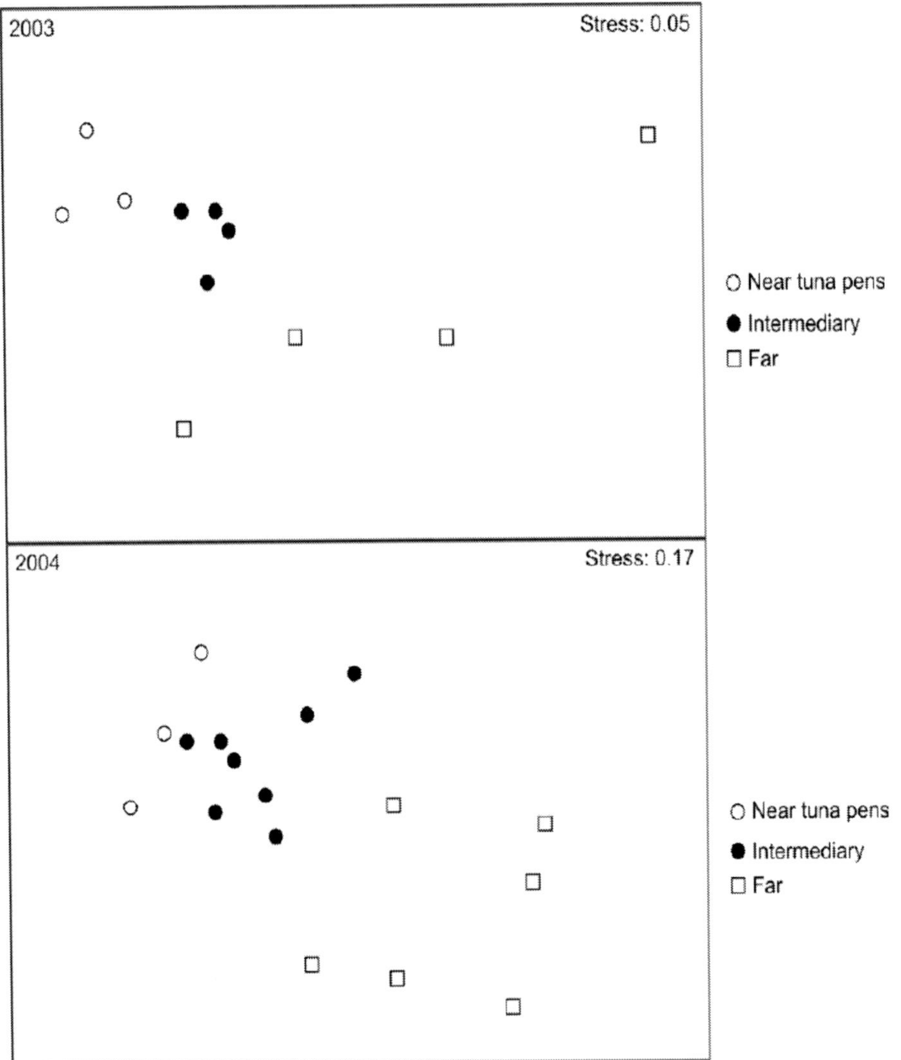

Figure 13. Non-metric multidimensional scaling (MDS) ordination of stations sampled at Salsipuedes, Baja California. Sampling sites were well classified together, based on their distances to the tuna pens.

Conclusions

In the selected study sites along the western coast of Baja California peninsula, macrofauna communities were characterized by a diverse composition. Polychaetes were the most representative taxon in terms of abundance and species richness.

In all the sites the dominant polychaetes species belong mostly to the families Spionidae, Capitellidae, Paraonidae, Cirratulidae, Maldanidae, Ampharetidae, Nephtyidae, Capitellidae, Onuphidae and Syllidae. Among the more specious families were consistently Cirratulidae, Spionidae and Paraonidae.

Food supply is a key factor structuring marine benthic communities (Pearson and Rosenberg, 1987; Wieking and Kröncke, 2005). Four trophic groups were found: filter feeders, deposit-feeders, herbivores and carnivores. Surface and subsurface deposit-feeders were dominant, mostly associated with muddy sediments rich in organic matter (presented higher densities in areas of low hydrodynamics), followed by carnivores (exploiting the abundance of opportunistic deposit-feeders) and suspension-feeders (which predominated in coarser sediments).

The family Spionidae was the most abundant, widespread and well represented polychaete family. This family can feed as a suspension-feeder or as a surface deposit-feeder depending on environmental conditions. At Salsipuedes trophic changes (increase of detritivore species) could be explained by the constant arrival of organic matter in this area.

Results showed that polychaetes abundance contributed to 45% (San Quintín), 50% (Magdalena), 62% (Salsipuedes) and 64% (Todos Santos) of the total abundance of the macrobenthic communities respectively, indicating there are a key component of the benthos.

Bahía Todos Santos had the highest abundances and Shannon diversity values. In this bay around one third of the study sites presented Shannon diversity values higher than 4.00, which are indicative of an extremely favorable environment for the development of benthic invertebrates and particularly annelids.

At Salsipuedes, the MDS analysis revealed that the similarity of polychaete community structure depended on the distance from tuna pens. Sampling sites were well classified together, based on their distances to the tuna pens. We detected a moderate impact from the tuna fattening probably because the nearest we could sample from tuna pens was 250 m, it is an open bay and hydrodynamics of the area are favorable.

Relevance

The oceans cover approximately 70% of the Earth and sediments cover most of the ocean floor, the organisms that reside in these sediments therefore constitute the largest faunal assemblage on Earth in aerial coverage (Snelgrove, 1998). The biomass in these sediments is dominated by macrofauna, a grouping of invertebrate polychaetes, molluscs, crustaceans and other phyla. Macrofauna in marine sediments also play important roles in ecosystem processes such as nutrient cycling, pollutant metabolism, dispersion, burial, and secondary production. Macrofaunal activity also impacts global carbon thus, evaluating the composition and structure of macrobenthic communities is of paramount importance to understand the functioning of benthic ecosystems. Marine invertebrates provide a critical prey resource for various types of sea birds (e.g. shorebirds), marine mammals (sea otters), marine fishes and larger invertebrates (e.g. cephalopods). Macrofauna enhance productivity in the area as they attract marine life. Many of these areas are nursery sites for diverse species. Some invertebrates can live several years and integrate environmental conditions which are important for monitoring studies. Some crustaceans, molluscs and polychaetes are of importance as seafood and aquaculture species (Mann, 1984; Olive, 1994). New species and genera have been found (León-González and Díaz-Castañeda. 1998; Díaz-Castañeda and San Martin. 2001). Some taxa are poorly known and certainly many remain undescribed. We hope to contribute to an increase in the knowledge regarding marine invertebrates in the Mexican Pacific.

Future Research

The rate at which human activities are changing marine ecosystems particularly in the coastal zone is alarming; many species are disappearing without previous documentation of their existence and with potential changes in ecosystem services provided. This could have important ecological and socio-economic consequences.

There is an important gap in our knowledge of the marine biodiversity of benthic communities in the Mexican Pacific, there are extensive coastal areas and only a small fraction of them has been studied. These kind of studies are laborious and time consuming and require trained scientists. Establishing base-line information is of vital importance in order to follow changes in marine ecosystems and define areas important for marine conservation (invertebrate larvae availability and renewal of benthic communities).

Comparisons of benthic community structure and ecosystem functioning along the eastern Mexican Pacific are scarce. The use of stable isotopic tracers will contribute to increase our understanding of the energy and food web relations in benthic communities (Peterson, 1999) but extensive work remains to be done.

Mid-water oxygen minima (<0.5 ml/l^{-1} dissolved O_2) intercept the continental margins along much of the eastern Pacific Ocean, creating extensive stretches of sea floor exposed to permanent, severe oxygen depletion. These seafloor oxygen minimum zones (OMZs) typically occur between 200 - 1000 m, and are major sites of carbon burial along the continental margins. OMZs generally form where strong upwelling leads to high surface productivity that sinks and degrades, depleting oxygen within the water column (Levin, 2003). The best-developed OMLs are found in the eastern tropical Pacific Ocean (Kamykowski and Zentara 1990). Off Mexico the OMZ is over 1000 m thick while off Chile, it is <400 m thick (Levin, 2003). Particulate organic carbon (POC) values of 3-6% are typical of OMZs (Levin and Gage, 1998), this represents an exceptional source of food for benthic organisms. We need to fill this gap in the knowledge of benthos in these areas studying the biological diversity, structure and adaptations of these benthic organisms. Besides, as global warming and eutrophication reduce oxygenation of the world ocean, there is a pressing need to understand the functional consequences of oxygen depletion in marine ecosystems (Levin, 2003).

Scleractinian Corals

Background

Even before the encounter between Europeans and native people, the Mexican coasts and waters from the Pacific had been studied. In his celebrated book about the conquering of Mexico, Diaz del Castillo (1568) described what he called "The Beasts' House" in the Sacred Temple of Tenochtitlán (now Mexico City), where the Aztecs had terrestrial and aquatic fauna from all over Mexico, Central and South America. He was impressed with the diversity of animals and the way Aztecs properly fed them. That was what we would call today a zoo and they existed here long before Leonardo da Vinci suggested their creation in Europe in the early XVI century. However, most of that traditional knowledge was lost mainly because of political reasons , the analysis of which is out of the scope of this chapter.

Verrill (1864, 1868) described the first species of hermatypic corals from the Mexican Pacific. One of the striking features about corals in this region was they do not resemble or have "sibling species" with species from the Caribbean, as most marine groups do, but they are more similar to species from the Central and Western Pacific. During the last century several expeditions continued to explore the Mexican Pacific, notably the Gulf of California (Durham, 1947; Durham and Barnard, 1952; Squires 1959) and ten new species were found. However, it was noticed that diversity was much lower that in the Mexican Caribbean (Villalobos, 1960).

Intense research was conducted during the last decades of the XX century, as can be acknowledged by the number of reported species (Figure 14). It is worth noting that during the period of the Mexican revolution, social and political conditions precluded research and there is another plateau during the 1960's and 1970's. Afterwards research has continued mostly by Mexican investigators rather than foreigners as before, notably Reyes-Bonilla *et al.* (2005).

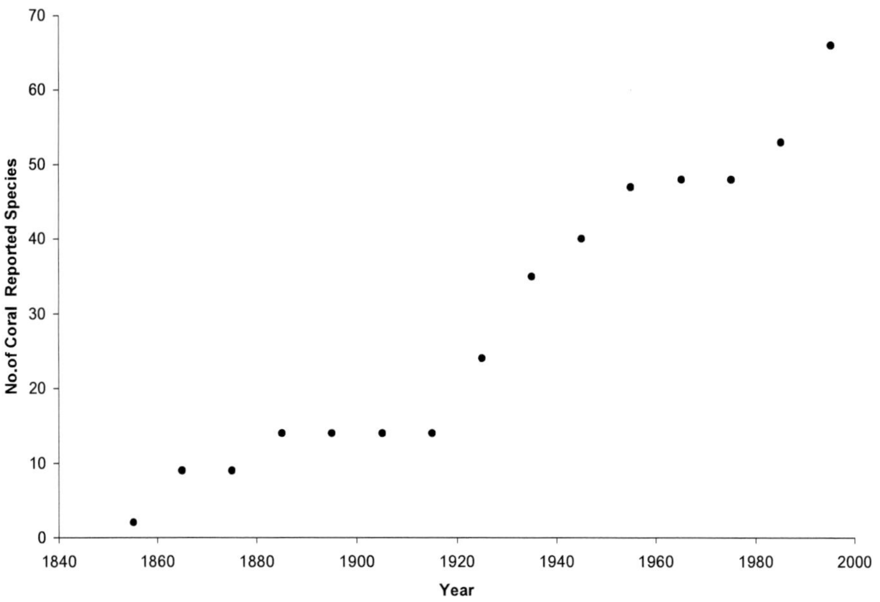

Figure 14. Number of reported species of coral from the Mexican Pacific from 1840 to 2005. Source: Reyes-Bonilla et al., 2005.

Methods and Source Data

This work is the result of a thorough review of all published data on corals from the Mexican Pacific, from the very first record (Verrill, 1864) to the last published in 2006. It also included visits to 17 collections (six in Mexico and 11 in the US; Table 5) in order to verify species identification and records.

The pattern of diversity at family level is a matter of distant geological history as seen in the fossil record, when the presence of the Tethys Sea and the Central American Seaway led to a uniform circumglobal fauna. The generic diversity is primarly created by the many taxa that evolved in, and remained in the Indo Pacific.

The pattern of diversity at species level is primarly the outcome of ocean circulation and resulting geographic and evolutionary changes from the Plio-Pleistocene to the present time (Veron 2000).

Web sites with records of coral species from the Mexican Pacific were also visited (University of Iowa http://www.uiowa.edu/; Australian Instiute of Marine Sciences http://www.aims.gov.au/; British Museum http://www.britishmuseum.org/). Field trips all along the coast were conducted between 2002 and 2005 and samples were collected if the species has not been reported before. This was done by SCUBA diving down to 30 m deep in 364 locations.

Brief Description of Main Oceanographic Features

Diversity of corals at the species levels is totally dependent on currents (see Figure 1). The paths of the currents are the paths of gene flow, since dispersal occurs during the larval stage. Therefore, understanding the ocean currents is essential to understand the distribution of the coral species.

The western coast of Mexico in the Eastern Pacific is a transition zone between tropical and temperate regions. It features very particular oceanographic conditions, shaped mainly by the confluence of the water masses of the Costa Rica Coastal Current (CRCC), the California Current (CC), and of the Gulf of California (GC). The CRCC and the CC converge off of Tehuantepec in March – April, but off of Baja California in September – October (Badan-Dangon 1998). In winter, the main flow of the CC does not reach the mouth of the GC; it turns eastward to join the North Equatorial Current. Northerly winds in the GC force circulation southwards, which continues along the coast. In the Gulf of Tehuantepec, strong northerly winds force large anticyclonic eddies that carry coastal waters to the Pacific at least to 110° W (Trasviña and Andrade 2002). During the summer, circulation turns northwards in the form of localized currents, which are occasionally forced by local winds. There is no evidence of a consistent coastal circulation such as the CRCC. Therefore, there is no exchange of physical properties between Costa Rica and southern Mexico (Trasviña and Andrade, op.cit).

The regional oceanography responds to particular phenomena and associated forcing in different spatial and temporal scales. The hydrodynamics on average (i.e. decades), are governed by the subtropical gyre of the North Pacific, and forced by long-term average of the wind pattern (Pares Sierra et al.1997). On an annual scale, variability is determined by El Niño, which is remotely forced as a result of its connection to the Equatorial Pacific. Seasonal variations are due to local wind forcing while more local events, such as upwelling events, jets and eddies, also force instability of the mean current (Pares Sierra *et al.*, op. cit).

Despite these complex hydrodynamics, it hosts some well-developed coral reefs, which harbor rich, diverse ecosystems (Durham 1947; Squires 1959; Reyes-Bonilla 2003).

Results

Specimens reviewed from collections and published data resulted in six taxa not previously recorded in Mexico: two zooxanthellate species: *Psammocora haimeana* Milne Edwards and Haime, 1851 and *Fungia vaughani* Boschma, 1923; and four azooxanthellate species: *Astrangia equatorialis* Durham and Barnard, 1952, *Labyrinthocyathus quaylei* (Durham, 1947), *Polymyces montereyensis* Durham, 1947 and *Tubastrea tagusensis* Wells, 1982. Two of these new records represent small range extension: *P. montereyensis* and *L. quaylei*, from California (Cairns, 1994; Reyes Bonilla and Cruz Piñon, 2000) but the other are far more important. *A. equatorialis* and *T. tagusensis* were thought to be endemic of Galapagos and Central America (Cairns, 1991; Reyes Bonilla, 2001) and were found in the coast of Jalisco (Cupul Magaña et al., 2000; USNM 96513), while *P. hamieana* has never been seen in Eastern Pacific (Glynn and Ault, 2000; Reyes Bonilla, 2001) but a specimen misidentified collected in Las Cruces, Baja California Sur (24° 12') in 1974 in the California Academy of Science collection (CASIZ 2304). *F. vaughani* had been collected in the Revillagigedo but its identification was not conclusive (Ketchum and Reyes Bonilla, 2001). However, some specimens collected in the Revillagigedo in different decades were found in collections (USNM and SBNHM) therefore confirming the presence of stable populations in the Revillagigedo Islands.

So far, 67 species of scleractinian corals, belonging to 12 families (Pocilloporidae, Poritidae, Agariciidae, Siderastreidae, Fungiidae, Fungiacyathidae, Micrabaciidae, Rhizangiidae, Caryophylliidae, Turbinoliidae, Flabellidae, Dendrophylliidae) have been reported from the 11 coastal states from the Mexican Pacific (Reyes- Bonilla et al., 2005; Figure 15). Baja California Sur has the largest number of reported species (35) and Chiapas the lowest with just one species of an azoxanthellate coral reported. However, Colima has more zooxanthellate corals than any other state. This is because the Revillagigedo Islands politically belong to this state and this location has the largest richness of all the Mexican Pacific and includes migrant species from the Indo Pacific such as Porites lutea, Pocillopora woodjonesi and Pavona maldivensis.

Cluster analysis of zoxanthellate corals shows that there are two major groups: one formed by Baja California, Sonora and Sinaloa, states bordering the Gulf of California and other group spliting Michoacan from the rest, followed by Baja California Sur and higher similarity among Nayarit, Colima, Jalisco, Guerrero and Oaxaca (Figure 16). This array is consistent with other taxonomic groups: the Gulf of California province (*sensu* Briggs, 1975; Brusca and Wallerstein, 1978, among others) and the Tropical Pacific or Mexican province *sensu* Briggs (1975). The Gulf of California has a unique fauna with some endemic species (e.g. *Porites sverdrupi*, *Astrangia cortezi*, *Ceratotrochus franciscana*). Moreover, this clustering show that several species which are migrants from the Western Pacific and that first appeared in the Revillagigedo's were able to colonize Nayarit and southwestern Gulf of California, such as *Pavona duerdeni*, *Psammocora superficialis* and *Fungia vaughani*. Baja California Sur has mixed species, sharing species such as *Pocillopora damicornis*, *Porites panamensis*, *Pavona gigantea* and therefore could be considered as a transition zone between the Gulf of California and the Tropical Pacific. Guerrero and Oaxaca have distinctive species *Pavona varians*, and *Pocillopora effusus* which most likely were carried by the CRCC from Central America (Reyes Bonilla 2003). In short, hermatypic corals from the Mexican Pacific have two biogeographic zones: North and East of the Gulf of California, where the fauna is monogeneric (*Porites*), and the Tropical Pacific, with a subdivision marked by Jalisco.

Figure 15. Coastal states of the Mexican Pacific. 1. Baja California (BC), 2 Baja California Sur (BCS), 3 Sonora (SON), 4 Sinaloa (SIN), 5 Nayarit (NAY), 6 Jalisco (JAL), 7 Colima (COL), 8 Michoacán (MIC), 9Guerrero (GRO), 10 Oaxaca (OAX), 11 Chiapas (CHI).

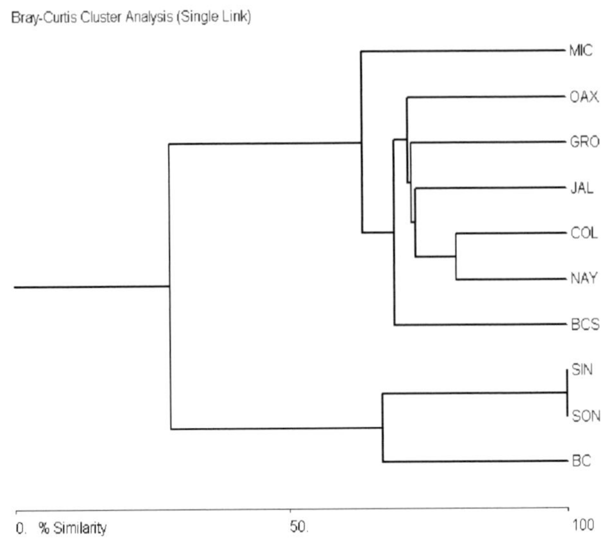

Figure 16. Dendrogram of similarity of zoxanthellate corals species composition in the Mexican Pacific. Abbreviations as in figure 15.

On the other hand, cluster analysis performed with azoxanthallate corals showed much less similariy and resembles a latitudinal gradient of similarity, excluding Colima which is the less alike regarding this species composition (Figure 17). Again, states bordering the Gulf of California (BC, BCS and SON) have distinct species such as *Astrangia californica, Dendrophyllia oldroydae* and *Balanophyllia cedrosensis*. On the other hand, there are very few species "typical" from the Tropical Pacific, e.g. *Astrangia browni, A. tangolaensis* or *Paracyathus humilis* and they occur isolate. In short, cluster analysis of azooxanthellate species clearly resembles a latitudinal gradient, as noted by Reyes Bonilla and Cruz Piñón (2000).

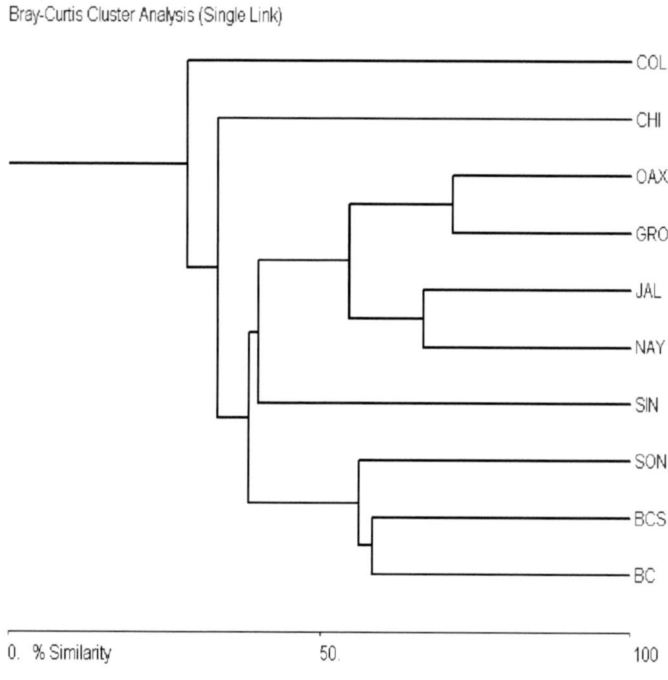

Figure 17. Dendrogram of similarity of azoxanthellate corals species composition in the Mexican Pacific. Abbreviations as in figure 15.

Relevance

Coral reefs are widely recognized as diverse and productive ecosystems. In spite of the harsh conditions imposed by cold water, upwelling, low pH and narrow continental shelf, there are real coral reefs in the Mexican Pacific, notably in the Revillagigedo Islands (18° 43'22"N, 110 ° 57' 5"W), Bahía de Banderas (25° 52′N, 105° 54′W), Bahías de Huatulco (15° 45'N, 96° 07 W) and Cabo Pulmo (23°50'N, 109°25'W). This work shows that the coral species richness is far larger than previously thought (e.g. Veron 2000) and that there might be more species to be discovered or reported, mainly from deep water.

Future Research

Genetic studies of corals in the Mexican Pacific are incipient. So far there are conflicting results about the center of diversity in the region. The natural candidate would be the Revillagigedo Islands, assuming migration from the Indo-Pacific. However, Saavedra-Sotelo (2007) found partial agreement between the degree of connectivity among localities and the general pattern of circulation of the region, suggesting the influence of historical and contemporary factors in the demographic connectivity between *Pavona gigantea* populations along the Mexican Pacific.

Assessing the vulnerability of corals from the Mexican Pacific to global change might be highly relevant, since they already live in low alkalinity waters. A preliminary work showed that carbonate deposition has been decreasing during the last two decades (Calderon-Aguilera et al. 2007) due to live coral cover loss.

Zooplankton in the Eastern Tropical Pacific

Background

Biogeographic studies have shown that the distributions of pelagic animals follow the patterns of large-scale circulation, characterized by the distribution of water masses (Steuer, 1933; Sverdrup, Johnson, and Fleming, 1942; McGowan, 1960a, b, 1974; Bogorov, 1958; Beklemishev, 1961, 1971, 1981; Johnson and Brinton, 1963; Semenov and Berman, 1977; Reid, Brinton, Fleming, and Venrick , 1978; Van der Spoel and Pierrot-Bults, 1979; and Van der Spoel and Heyman, 1983). The eastern tropical region of the Pacific has been considered a biogeographically, distinct province (Ebeling, 1967; McGowan, 1971, 1974, 1986; Briggs, 1975; Vermeij, 1978; Hastings, 2000).

In this area we have the Oxygen Minimum Zone (OMZ), which is very shallow, almost anoxic, layer up to 1,000m thick occurring at shallower depths than other low-oxygen areas. In Figure 18 we show the mean oxygen minimum depth using 1ml/l as the threshold value. It is evident that the Mexican tropical Pacific has a very shallow OMZ.

Isolated efforts to study this area, such as the EASTROPAC program, have taken place, however only few taxonomic groups have been studied (Love, 1971; Ahlstrom, 1971, 1972) including ichtyoplakton, or euphausiids. Low oxygen may constitute a barrier to zooplankton and thus vertical migration (VM) could be affected. Due to this lack, or restriction, of vertical migration, the trophic relationships may be altered, and some organisms, could, and actually do, adapt to these conditions (Antezana, 2002a, 2002b, Childress and Seibel, 1998; Fernandez-Álamo and Färber-Lorda, 2006), making this area very interesting to study. Some animals implement anaerobic metabolism to support these conditions and even survive in anoxic waters for some time (Childress and Seibel, 1998), which is a very useful adaptation for vertical migrating organisms. However more studies are necessary to understand zooplankton adaptations to low oxygen.

Mesoscale

In the eastern tropical Pacific, the mesoscale features are mostly associated with the wind jets blowing through three gaps in México and the Central American mountain range: at the Tehuantepec isthmus in Mexico, the Gulf of Papagayo region of Nicaragua, and the Panama Canal (Amador et al., 2006). This produces jets that force strong mixing and upwelling (Kessler, 2006; Willett et al, 2006) and is reflected in high biological productivity in the area (Färber-Lorda et al., 1994, 2004b; Gonzalez-Silvera et al., 2004; Pennington et al., 2006).

In the Eastern tropical Pacific (ETPac), iron enrichment experiments have shown ambiguous responses by zooplankton (Landry et al., 1997; Landry et al., 2000a; Landry et al., 2000b), with an increase of zooplankton biomasses, followed by a rapid return to previous conditions. This could be related to the presence of two trophic pathways or food webs, in tropical areas. One is the "microbial loop" in which bacteria are eaten by microzooplankton which are then eaten by, mesozooplankton , with microzooplankton may constituting up to 80% of food for mesozooplankton, instead of the usual phytoplankton. The other is the "classic" food chain in which mesozooplankton feeds mostly on phytoplankton. This more complicated food web with two co-existing food pathways may regulate the secondary productivity in an efficient way, but this is just an hypothesis. More studies are necessary to understand these mechanisms.

In general, small seasonal changes exist in the eastern tropical Pacific, which includes the Mexican tropical Pacific. Fernandez-Álamo and Färber-Lorda, (2006) show local conditions are more important than seasonal changes, and that coastal upwelling is a decisive factor determined mostly by dominant forcing winds, such as the conditions found in autumn and winter in the Gulf of Tehuantepec (Färber-Lorda et al., 1994 and Färber-Lorda et al., 2004b), or by the seasonal joining of the West Mexican Current and the remains of the Costa Rica Coastal Current (Färber-Lorda et al., 2004c).

Given the importance of OMZ areas in the world oceans, this review will propose new research and ideas in the study of zooplankton in the ETPac and especially in the Tropical Mexican Pacific. In this part we will concentrate on the Tropical Mexican Pacific, because more studies are needed in this area. The main purpose of this contribution is to draw attention to its importance, and the implications of the already mentioned different conditions.

General Methods

Zooplankton sampling was performed with different nets, with different mesh sizes and mouth sizes, with multiple sampling nets utilized. The Longhurst-Hardy type net was very useful to understand zooplankton vertical distribution before the invention of the multiple MOCNESS type net. Most of the historical data, obtained from the COPEPOD data set of the National Marine Fisheries Service, were obtained using a 333μm or 500μm mesh of a bongo net. However, given the fact that zooplankton in tropical waters are smaller, 333 μm data will be discussed here. Relationships with the physical, chemical and trophic environments will also be discussed, with some of the chemical methods described in Färber-Lorda (2004c).

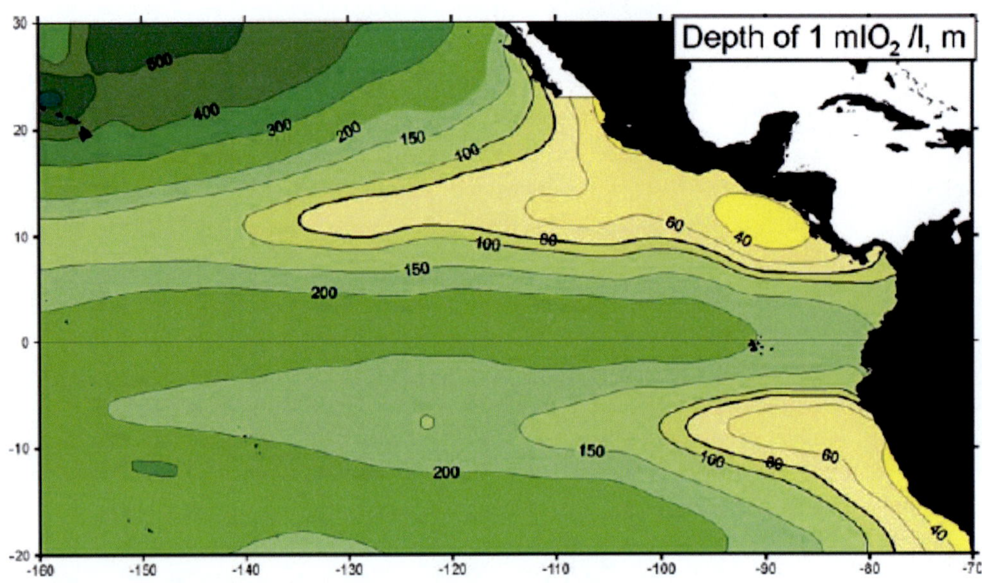

Figure 18. Mean depth of the oxygen minimum, considering the threshold as 1 ml O_2/ l. The entire Mexican coast is placed over a low and shallow oxygen minimum area (from: Fernàndez-Alamo and Färber-Lorda, 2006).

Results

Spatial and Temporal Patterns

As previously stated (Fernandez-Àlamo and Färber-Lorda, 2006) we know that zooplankton productivity in the tropics depends mostly on local conditions, such as upwelling and wind regimes, and that seasonal differences are not as important as those in temperate and polar seas. These patterns follow the general hydrography of the area, and are in agreement with major currents, surface water masses, and upwelling regions (Brandhorst, 1958; Holmes, 1958; Reid, 1962; and Forsbergh and Reid, 1964). Some of these papers also show the relationships between zooplankton biomass and chemicals, such as nutrients and chlorophyll, and biological data, such as phytoplankton and micronekton, birds, mammals and fisheries. In Figure 19, the general distribution of zooplankton in the Easter tropical Pacific is shown (from Fernandez-Àlamo and Färber-Lorda, 2006). This figure was plotted with historical data (COPEPOD Data), mostly from the EASTROPAC cruises. It clearly shows a higher zooplankton volume around the Costa Rican dome and the Gulf of Tehuantepec, as well as in the Humboldt Current in front of Peru, Ecuador and the North of Chile, and a long tongue-like area around the equator.

Between 1952 and 1964, 29 biological-oceanographic expeditions were performed in the area, and the main results were reviewed by Blackburn (1966a), mostly on biological oceanography. A description of the oceanic circulation was the most important result. The only survey in this region that studied a large area, repeatedly, over a year (seven bimonthly periods) in the eastern tropical Pacific was the multi-institutional and international project EASTROPAC in 1967-1968, during which the tropical Mexican Pacific was partially studied.

Among the most important results of this survey is Love's (1972-1978) edition and publication of atlases that included physical, chemical, and biological data.

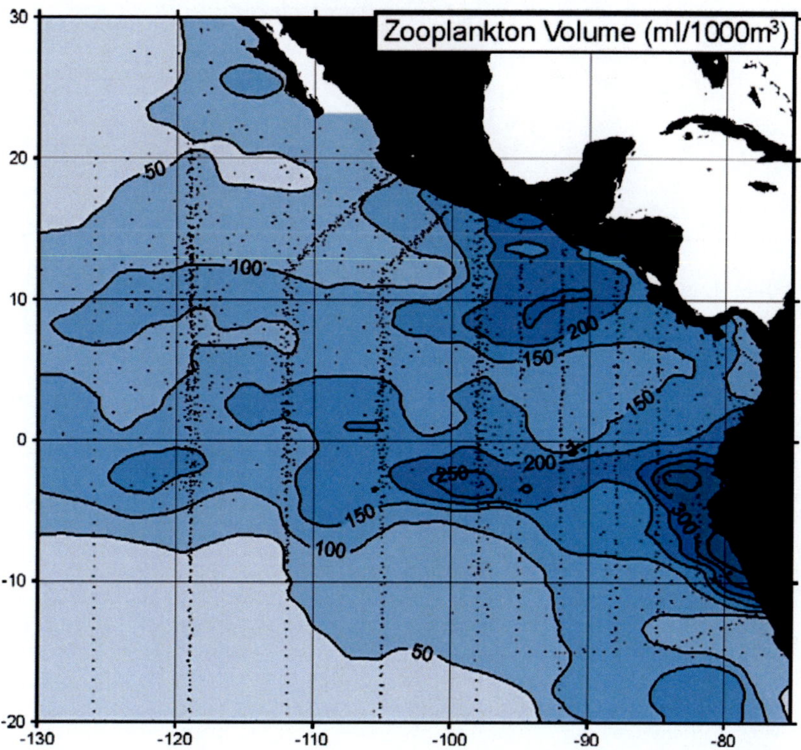

Figure 19. Mean zooplankton volume in the eastern tropical Pacific from bongo net, 333-μm mesh tows from 200-0 m (N=2405). Highest values are found around the Costa Rica Dome and the Gulf of Tehuantepec in the northern hemisphere and in front of Peru and Ecuador in the Southern hemisphere.

Many scientists have utilized the great amount of data collected and they have examined and analyzed the samples which resulted in many publications about different subjects. During these expeditions, a prospecting of commercial species was performed, in relation to areas of high abundance of plankton in upwelling regions. Only a few studies have been performed with multiple nets. Saltzman and Wishner (1997a) proposed a pattern of vertical distribution of the main groups of zooplankton in the OMZ in which they found 2 peaks of zooplankton, one on the thermocline and another just under the lower bottom of the OMZ layer. The question that remains is how these animals got there? But other more important ones, such as: What are the adaptations of these animals to live within the OMZ?

The dominant euphausiid species *Euphausia eximia* and *Nyctiphanes simplex* were found associated with the crustacean galatheid *Pleuroncodes planipes* by Brinton (1979), and are found in the transition zone between the California Current and the Eastern Tropical Pacific (the first species being a neritic species). This author utilized samples taken along a transect from 23° N to 3° S and studied the physical-chemical environment influencing the distribution patterns and vertical migration of euphausiids. In the region of extreme O_2 deficiency from 21° to 10° N, the dominant species are *Stylocheiron affine*, *S. carinatum*, *Euphausia tenera*, *E. distinguenda* and *E. lamelligera*. These last two species are considered the characteristic euphausiid species of the oxygen deficient waters of the tropical Mexican Pacific.

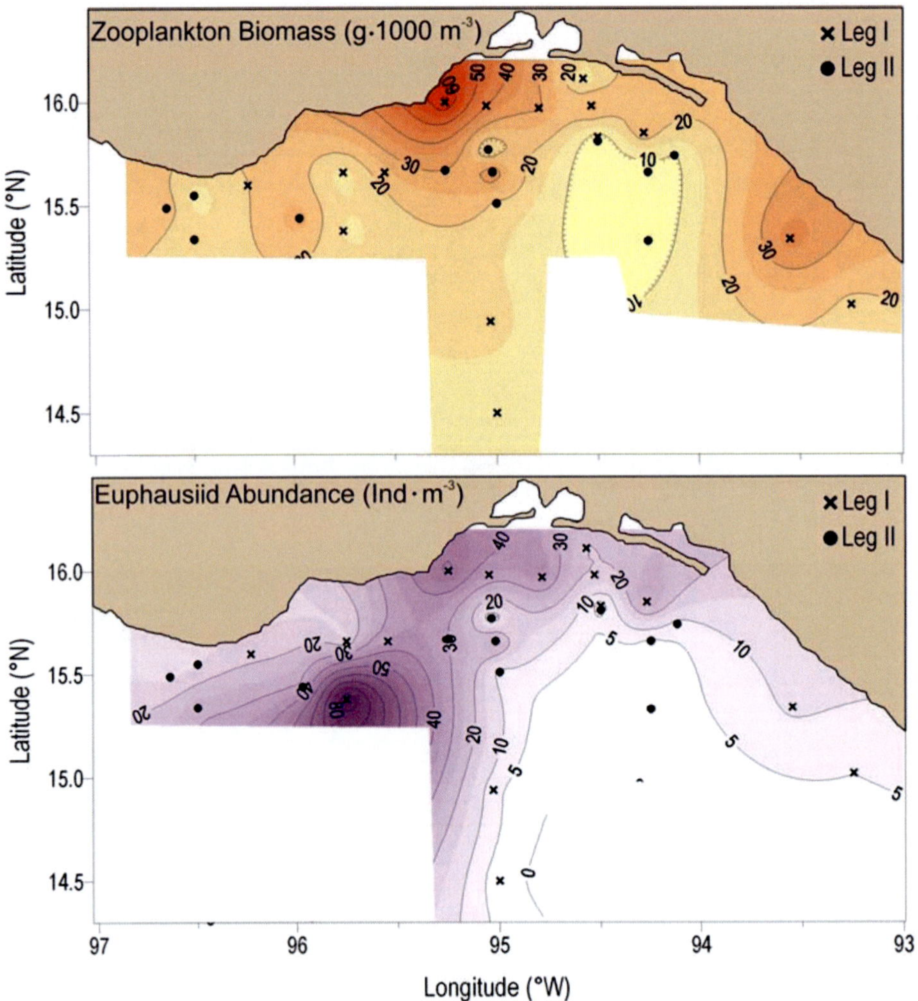

Figure 20a. Zooplankton abundance and Euphausiid abundance during intermitent wind forcing in the Gulf of Tehuantepec. High values were found in the mid-Gulf area. Leg 1, is the period before a wind event and Leg 2 after a wind event.

Recently, few studies have been performed in the tropical Mexican Pacific waters (Acal, 1991; Färber-Lorda, et al. 1994a, Färber-Lorda, et al., 2004b; Färber-Lorda, et al. 2004c; Franco-Gordo et al. 2001; Franco-Gordo et al., 1999, Franco-Gordo et al. 2004), but they are isolated, without continuity, and more sampling in the area is badly needed. In the Gulf of Tehuantepec, some papers have been published on the distribution of different zooplankton taxa, such as Alameda de la Mora (1980), and Fernandez-Àlamo (2000) for copepods, medusae by Segura Puertas(1987), Lopèz- Cortès (1990) Färber-Lorda et al. (1994) for euphausiids, and for holoplanktonic polychetes by Fernandez-Àlamos (1987). Acal (1991) studied the ichthyoplankton of the Mexican tropical Pacific, and found 40 families and 129 species. The most abundant species were *Bregmaceros bathymaster* and *Viciguerria lucetia*, with higher diversity found during the night related to vertical migration. Franco-Gordo et al. 1999 and 2001 described the seasonal zooplankton biomass and ichthyoplankton especies in the coasts of Colima and Jalisco. Franco-Gordo et al., (2004), described the trophic environment before and during the 1997-1998 ENSO, and found higher phytoplankton during El Niño, with small diatoms as the dominant component. The highest biomasses were found at coastal stations during the pre-ENSO period, with two periods clearly defined - a pre-ENSO (December 1995-March 1997) and an ENSO period (July 1997-September 1998).

In the Gulf of Tehuantepec, Färber-Lorda et al. (1994a) found that the neritic euphausiid *Euphausia lamelligera* was the dominant species (91% of the total abundance) during wind forcing conditions. Sampling was performed in coastal waters under the influence of strong intermittent winds, from land to sea, which produces strong vertical mixing and a gyre on the Western side of the Gulf, which results, also, in improved trophic conditions (Färber-Lorda, et al. 2004b). In Figure 20a we can see the zooplankton biomass and euphausiid abundance isolines, which show higher values of both variables in the mid-Gulf, where there is a strong influence of the winds, and when hydrographic conditions changed considerably in relation to wind stress.

Figure 20b shows the difference in surface temperature, which is lower in the mid-Gulf after a wind event (leg 2), as compared to before the wind event (leg 1). In the entrance of the Sea of Cortes, Färber-Lorda et al. (submitted) found again a dominance of *Euphausia lamelligera*, but the abundances were considerably lower, and this species was present only in a few near-shore stations, which is what was expected for a neritic species. Other important species in the area were *E. eximia* and *E. tenera*, present in stations closer to the California Current waters and *E. distinguenda* in the oceanic tropical waters. A coincidence, though not significant, of higher abundance with higher POM (in this case particulate protein+particulate carbohydrates) was found in the area (Färber Lorda et al. 2004c).

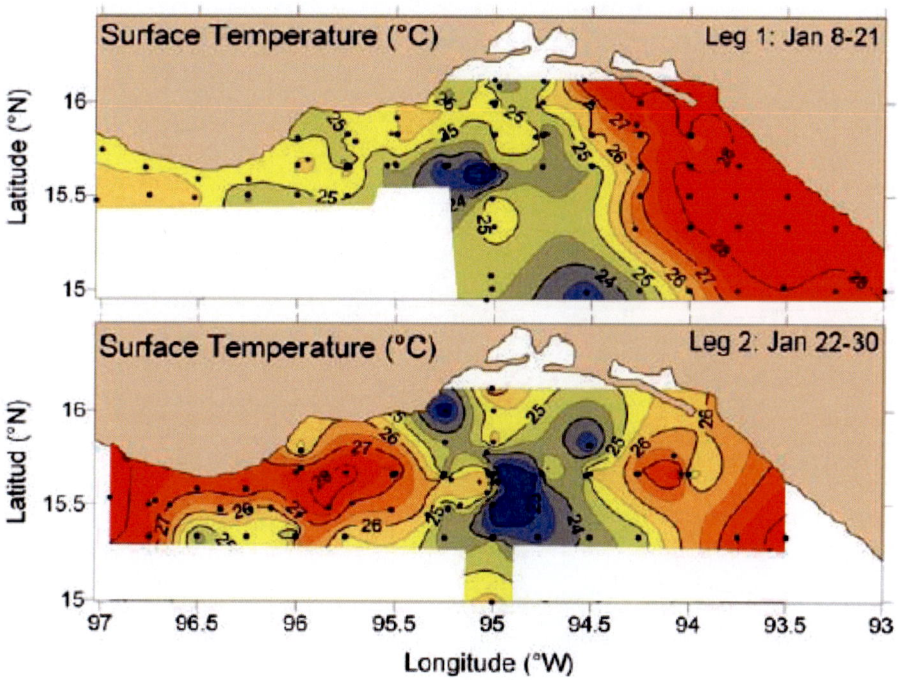

Figure 20b. Surface temperature in the gulf of Tehuantepec before (Leg 1) and after (Leg 2) a wind event, lower temperature occupied a wider area in the mid-Gulf, and apparently two gyres were produced at both sides of the central wind jet.

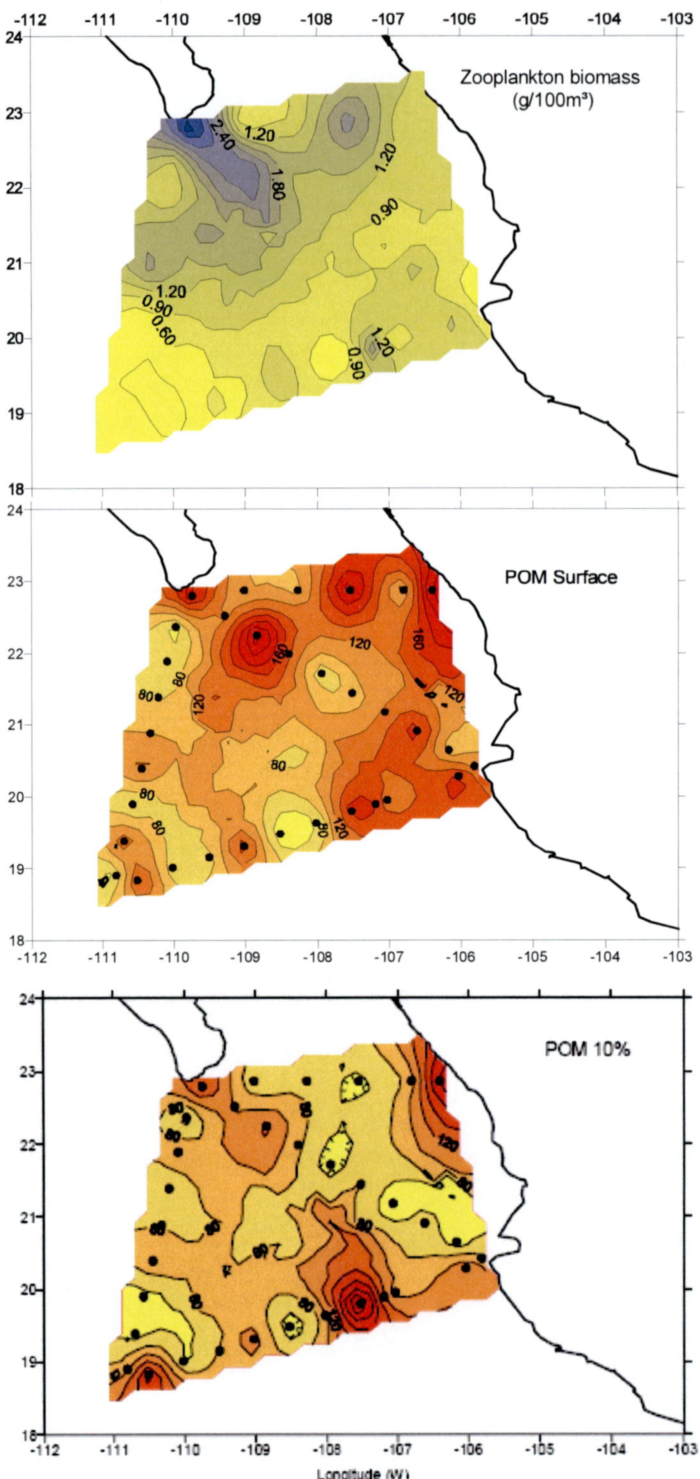

Figure 21. Zooplankton biomass, POM (particulate protein + particulate carbohydrates), in the entrance of the Sea of Cortés. A coincidence between high zooplankton biomasses and high POM was found.

In the same area Färber Lorda et al. (2004c) found a significant regression between zooplankton biomass and POM, for three different hydrographic areas obtained with their T-S diagrams. In Figure 21 we show the distribution of biomass and POM at different light attenuations (modified from Färber Lorda et al., 2004c), showing the mentioned coincidence of high zooplankton biomasses with high POM.

Relevance

All these studies and papers have shown that zooplankton distribution is controlled by the main features of the hydrographic conditions, and by the dominant winds of the area. High productivity areas in the Mexican Tropical Pacific are mostly restricted to upwelling areas, which are caused mainly by wind driven circulation and vertical mixing. Around capes like Cabo Corrientes and Cabo San Lucas, high productivity is also found, mainly due to water masses mixing, such as the area in front of Cabo San Lucas, which is caused by the mixing of California Current waters with the west Mexican Current and the Gulf of California waters. As mentioned, seasonal changes are not important in this area, except for the dominant wind driven circulation, which can change primary and secondary productivity. We now know that some species are well adapted (Antezana, 2002b) to low oxygen, which is a useful adaptation for this area. The absence of a continental shelf is another factor in the relatively low productivity of the area, as compared to the low oxygen area of the Humboldt Current. Greater and more frequent sampling is necessary in the area to understand the seasonal changes. More studies on the adaptations of the species to low oxygen are necessary for micro, meso and macrozooplankton, as well as the importance of the "microbial loop" as food for microzooplankton, and the importance of microzooplankton as food for mesozooplankton within the OMZ.

Future Research

The Eastern Tropical Pacific is a unique area in the world oceans, and its study is a great challenge. A carbon budget more dependent on bacteria and microzooplankton has developed in the eastern tropical Pacific, together with the classic food chain, resulting in a more complex food web. Productivity in the ETPac is regulated by a more complex trophic structure, as compare to other oceans. In Table 7 we present important aspects that need to be studied.

Some of the important questions and challenges that arise in this huge area are: What are the mechanisms by which such a wide oxygen minimum area has developed? What are the consequences of nutrient dynamics in the area? Which is the real importance of microzooplankton and bacteria in the area, and their trophic interactions? How does the OMZ affect bacterial biomass and microzooplankton feeding on them? Why do iron enrichment experiments in the area show such an ambiguous response with a fast return to previous conditions? Which are the adaptive mechanisms of zooplankton living in low oxygen areas? in particular in the ETPac? What are the adaptations of the different groups of zooplankton that live in the OMZ, and which are these groups? By exploring these above questions, we can acquire better knowledge of zooplankton vertical distribution and vertical migration, and its relation to trophic and physical conditions in this area. This will lead to a greater understanding for the importance for benthic communities, carbon recycling for bacteria, and for the "microbial loop".

Table 7. Vertical distribution of zooplankton in a typical area of the OMZ. Two zooplankton peaks were found. This figure was made to remark the main questions still unanswered for a better understanding of the ETPac and especially the Mexican tropical Pacific.

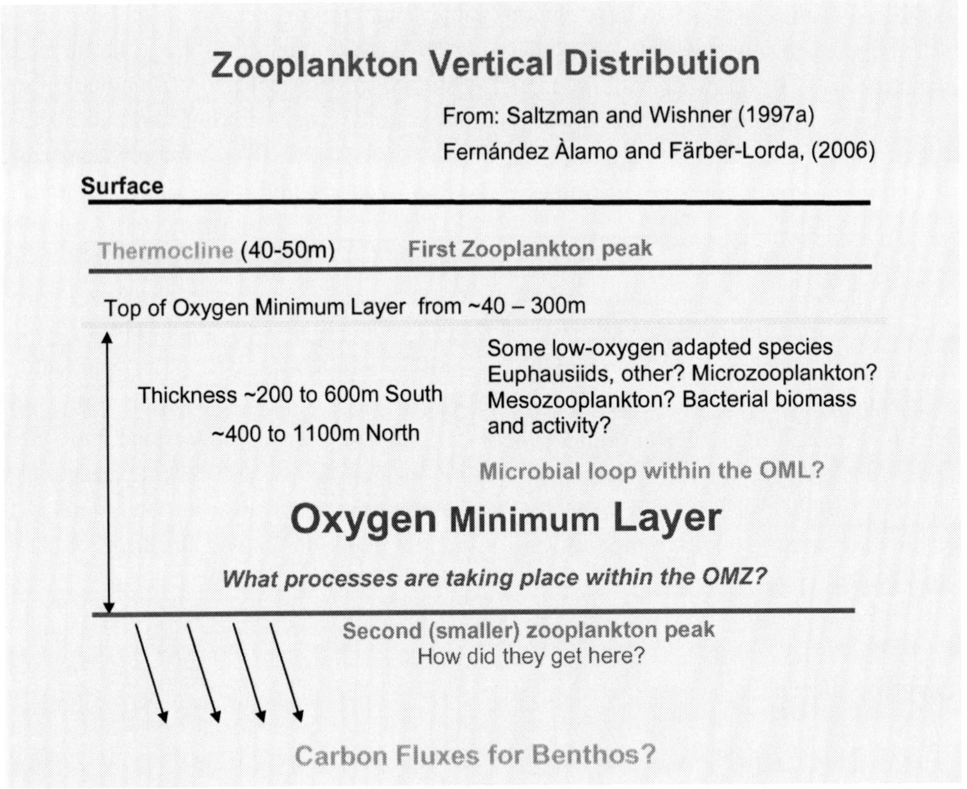

SYNTHESIS

This chapter on benthic and pelagic communities in the Mexican Pacific shows some of the important advances in the region as well as many gaps that need to be addressed. In general we acknowledge an impressive productivity and diversity in the area and note that the physical environment allows the flourishing of important biological communities: euphasids and other planktonic organisms, macrobenthos and particularly polychaetes, crustaceans and molluscs, a great diversity of macroalgae and rocky-shore invertebrates as well as coral communities.

As we have seen the kelp forest ecosystem provides food, habitat, and refuge for many coastal temperate marine organisms (many with economic importance) and is highly susceptible to climate change and ocean warming. It is not well understood how temperate kelp populations can survive in central and southern Baja California at their southern limit where water temperatures are seasonally warm and how the populations can recover rapidly after severe disturbance caused by El Niño events and this is an important point for further study. Although in Northern Baja California (Ensenada and Punta Baja), the temperature-nutrient relationship behaves in a similar way to that established for California, with a lack of nutrients in water warmer than 15°C, waters further south (Abreojos or San Juanico) show higher nutrients in warmer waters and it is still unclear what mechanisms control this pattern. Transport of nutrients, contaminants, and larvae often result from the same physical mechanisms. Therefore, transport mechanisms for invertebrate larvae are also linked to high frequency mechanisms and have revealed interesting new results using barnacle cyprids as proxies. Understanding coastal zone fluxes of nutrients and larvae are important for the replenishment and maintenance of many kelp and invertebrate populations near their southern limit in this biogeographic transition zone.

Polychaetes play a major role in recycling marine sediments and the burial of organic matter. Processes occurring in marine sediments are controlled by a number of factors; among them are organic matter supply and bioturbation. The role of bioturbation is important because the infauna creates a three-dimensional mosaic of oxic and anoxic interfaces in sediments. The volume of oxic burrow walls (polychaetes, amphipods) can be many times the volume of oxic surface sediment thus modifying geochemical properties and allowing the arrival of more sensible species. In particular polychaetes are useful as indicators of disturbance/recovery as they respond exhibiting quantitative changes in assemblage distribution. The base-line information presented on polychaete communities of 4 different bays may serve as a reference in future studies, allowing comparisons and changes in biodiversity. Data from the selected sites: Todos Santos, San Quintín, Magdalena and Salsipuedes showed a diverse and abundant macrofauna, in which polychaetes were the dominant taxons and surface and subsurface deposit-feeders the dominant trophic category. Bahia Todos Santos presented the higher number of Phyla and diversity (Shannon index). San Quintin also presented an important diversity although after the ENSO 1997-98 there was a significant diminution of species richness and abundance. Bahía Magdalena the bigger bay in the western Mexican Pacific presented an important diversity; in 1999 after ENSO we also detected a decrease in species richness and diversity. At Salsipuedes, our results showed that some macrofaunal groups such as crustacean decapods and peracarids, echinoderms and annelids have increased their abundances, certainly exploiting the abundant organic matter which is accessible and comes from the tuna fattening facilities. Polychaetes contributed to 45-64% of the total abundance of macrobenthic communities indicating they are a key component of the benthos. The extense OMZ in the Mexican Pacific have not been studied yet; in other OMZs they have found high densities and low species richness for the benthic communities. The macrofauna within central Chile's OMZ consisted mainly of polychaetes of the families Cossuridae, Capitellidae, Spionidae, and Cirratulidae, although spionids often predominated when oxygen values ranged between 0.1 and 0.5 ml/L. On the slope beneath the OMZ, the dominant forms were polychaetes of the families Paraonidae, Amphinomidae, and Maldanidae (Gallardo et al., 2004).

The coral section is a review of all published data on corals from the Mexican Pacific, from the very first record (Verrill, 1864) to the last published in 2006. Corals in this area particularly at their northern distributional limits deal with complex hydrodynamics (upwelling, low pH) and variability patterns (e.g. ENSO) and have adapted and established important communities. Up to now 67 species of scleractinian corals and 12 families have been reported from eleven coastal states from the Mexican Pacific. This is the first biogeographical study that covers most of the Mexican Pacific. Corals are obviously important organisms forming the base of tropical ecosystems and are important indicators of contamination and climate change. Further study in these important key species, especially in understudied areas of the Mexican Pacific, is crucial to understanding diversity and determining how it may be threatened.

Zooplankton productivity in the tropics seem to depend mostly on local conditions as upwelling and wind regimes, seasonal differences are not as important as in temperate and polar regions. The dominant eupashiid species *Euphausia eximia* and *Nyctiphanes simplex*, were found associated with the crustacean galatheid *Pleuroncodes planipes* and are found in the transition zone between the California Current and the Eastern Tropical Pacific. In contrast in OMZs the dominant species are *Stylocheiron affine*, *S. carinatum*, *Euphausia tenera*, *E. distinguenda* and *E. lamelligera*. These last two species are considered the characteristic euphausiid species of the oxygen deficient waters of the tropical Mexican Pacific. This information can help monitor global warming.

One of the objectives of this paper was to contribute to establish an up-to-date review of our knowledge on some benthic (kelp, polychaetes) and zooplanktonic communities in this region, where the Mexican Pacific presents unique environmental characteristics for plankton as well as for higher organisms. This area is extremely rich and sustains many important fisheries, making it that much more relevant for further studies that will hopefully lead to a greater understanding and protection of the ecosystems of this region.

REFERENCES

Acal, E. D. (1991). Abundancia y diversidad del ictioplancton en el Pacífico Centro de México, Abril 1981. *Ciencias Marinas*, 17, 25-50.

Ahlstrom, E. H. (1971). Kinds and abundance of fish larvae in the eastern tropical Pacific, based on collections made on EASTROPAC I. *Fishery Bulletin*, 69, 3-77.

Ahlstrom, E. H. (1972a). Distributional Atlas of fish larvae in the California Current region: six common mesopelagic fishes-*Vinciguerria lucetia, Thiphoturus mexicanus, Stenobrachius leucopsarus, Leuroglossus stilbius, Bathylagus wesethi* and *Bathylagus ochotensis*, 1955 through 1960. *California Cooperative Oceanic Fisheries Investigations, Atlas* 17, 306 pp.

Ahlstrom, E. H. (1972b). Kinds and abundance of fish larvae in the eastern tropical Pacific on the second multivessel EASTROPAC survey, and observations on the annual cycle of larval abundance. *Fishery Bulletin*, 70, 1153-1242.

Ajmal-Khan, S., P. Murugesan, P. Lyla and S. Jaganathan. 2004. A new indicator macroinvertebrate of pollution and utility of graphical tools and diversity indices in pollution monitoring studies. Current Science, Vol. 87: 1508-1510.

Alameda de la Mora, G. (1980). *Sistemática y distribución de los copepodos (Crustacea) del Golfo de Tehuantepec, México*. Tesis de Licenciatura. Facultad de Ciencias, Universidad Nacional Autónoma de México. 280 pp.

Aller, R. C. 1978. The effects of animal-sediment interactions on geochemical processes near the sediment-water interface. In *Estuarine Processes,* (M. Wiley, ed.), pp. 157-172, Academic Press, New York.

Aller, R. C. 1982. The effects of macrobenthos on chemical properties of marine sediments and overlying water. In *Animal-sediment relations* (P. L. McCall and M. J. S. Tevesz, eds), pp. 53-102, Plenum Press, New York.

Álvarez-Borrego S, L. Galindo-Bect and B Chee-Barragán. 1975. Características hidroquímicas de Bahía Magdalena, B.C.S., México. Ciencias Marinas 2(2): 94-110.

Alvarez-Borrego S. and A. Chee-Barragan. 1976. Distribución superficial de fosfatos y silicatos en Bahía de San Quintín, B.C., *Ciencias Marinas* 3:51–61.

Alvarez-Borrego S., Ballesteros G. and A. Chee-Barragan. 1975. Estudio de algunas variables fisiciquímicas en Bahía de San Quintín, en verano, otoño e invierno. *Ciencias Marinas,* 2:1–9.

Amador, J. A., Alfaro, E. J., Lizano, O. G., and Magaña, V. O. (2004). Atmospheric forcing of the eastern tropical Pacific: A review. *Progress in Oceanography*, 69.101-142.

Antezana, T. (2002a) Vertical distribution and diel migration of *Euphausia mucronata* in the oxygen minimum layer of the Humboldt Current. In J. Färber-Lorda (ed.), *Oceanography of the Eastern Pacific. Volume II,* (pp.13-28) CICESE, Ensenada, México.

Antezana, T. (2002b) Adaptive behaviour of *Euphausia mucronata* in relation to the oxygen minimum layer of the Humboldt Current. In J. Färber-Lorda (ed.), *Oceanography of the Eastern Pacific. Volume II* (pp 29-40), CICESE, Ensenada, México.

Auad, G., Miller, A., and Di Lorenzo, E. 2006. Long-term forecast of oceanic conditions off California and their biological implications. *J. Geophysical Rese*arch, 111 .C9.

Badan-Dangon A 1998. *Coastal circulation from the Galapagos to the Gulf of California.* In: Robinson, AR,Brink KH (eds) The Sea. John Wiley and Sons, pp 315 – 343.

Baines, P.G. 1997. A fractal world of cloistered waves. *Nature* 388: 518-519.

Bakun, A. 1996. Patterns in the Ocean. Ocean Processes and Marine Population Dynamics. CIBNOR, Baja California Sur. 323 pp.

Barton, E. D., Argote, M. L., Brown, J., Kosro, P. M., Lavin, M., Robles, J. M., Smith, R. L., Trasviña, A., and Velez, H. S. (1993). Supersquirt: Dynamics of the Gulf of Tehuantepec, Mexico. *Oceanography*, 6, 23-30.

Bassin, C., Washburn, L., Brzezinski, M., and McPhee-Shaw, E. 2005. Sub-mesoscale coastal eddies observed by high frequency radar: A new mechanism for delivering nutrients to kelp forests in the Southern California Bight. *Geophysical Research Letters* 32: 1-4.

Beklemishev, C. W. (1961) On the spatial structure of plankton communities in dependence of type of oceanic circulation. Boundaries of ranges of oceanic plankton animals in the North Pacific. *Okeanologia*, 6, 1059-1072.

Beklemishev, C. W. (1971). *Distribution of plankton as related to micropaleontology.* In B.M. Funnel, and W.R. Riedel, *Micropaleontology of oceans*, (pp 75-87). London: Cambridge University Press.

Beklemishev, C. W. (1981). Biological structure of the Pacific Ocean as compared with two other oceans. *Journal of Plankton Research*, 3, 531-549.

Bilyard, S.C. 1987. The value of benthic infauna in marine pollution monitoring studies. Mar. Poll. Bull. Vol. 18: 581-585.

Blackburn, M. 1966a. Tuna oceanopgraphy in the eastern tropical Pacific.U. S. Fish and Wildlife Service, Special Scientific Report Fisheries 400, 1-48.

Bogorov, B. G. (1958). Biogeographical regions of the plankton of the North-Western Pacific Ocean and their influence on the deep sea. *Deep-Sea Research*, 5, 149-161.

Boström, C, E. Bonsdorff, P. Kangas and A. Norkko. 2002. Long-term changes of a brackish-water eelgrass (*Zostera marina* L.) community indicate effects of coastal eutrophication. *Estuarine, Coastal and Shelf Science* 55: 795-804.

Brandhorst, W. (1958). Thermocline topography, zooplankton standing crop, and mechanisms of fertilization in the eastern tropical Pacific. *Journal du Conseil Permanent International pour l'Exploration de la Mer*, 24, 16-31.

Briggs, J. C. (1975). *Marine Zoogeography*. New York: McGraw-Hill. 475 pp.

Briggs, J.C. 1975. *Global biogeography*. Elsevier, Amsterdam.

Brinton, E. (1979). Parameters relating to the distribution of planktonic organisms, especially Euphausiids in the eastern tropical Pacific. *Progress in Oceanography*, 8, 125-189.

Brusca, R.C. y B.R. Wallerstein. 1978. Zoogeographic patterns of idioteid isopods in the northeast Pacific, with a review of shallow-water zoogeography of the area. *Proceedings of the Biological Society of Washington* 3: 67-105.

Byers, S.C., Mills, E., Stewart, P., 1978. A comparison of methods to determine organic carbon in marine sediments with suggestions for a standard method. *Hydrobiologia* 58, 43–47.

Cairns, S.D. 1991. A revision of the ahermatypic Scleractinia of the Galápagos and Cocos Islands. *Smithsonian Contributions to Zoology* 504: 1-30.

Cairns, S.D. 1994. Scleractinia of the temperate North Pacific. *Smithsonian Contributions to Zoology* 557: 1-150.

Calderón Aguilera L. E., H. Reyes Bonilla y J.D. Carriquiry. 2007. El papel de los arrecifes coralinos en el flujo de carbono en el océano: estudios en el Pacífico mexicano. En: B. Hernández de la Torre y G. Gaxiola Castro (Eds.). Carbono en ecosistemas acuáticos de México. SEMARNAT – INE – CICESE. ISBN: 978-968-817-855-3. pp. 215 – 226.

Childress, J. J., and Seibel, B. 1998 Life at stable low oxygen levels: Adaptations of animals to oceanic oxygen minimum layers. The Journal of Experimental Biology 201, 1223-1232.

Clarke, K. 1993. Non-parametric multivariate analyses of changes in community structure. Australian Journal of Ecology 18: 117–143.

Cupul-Magaña ,A.C., O.S. Aranda-Mena, P. Medina-Rosas y V. Vizcaíno-Ochoa. 2000. Comunidades coralinas de las Islas Marietas, Bahía de Banderas, Jalisco-Nayarit, México. Mexicoa. 2: 15-22.

Dauer, D. 1993. *Biological Criteria, Environmental Health and Estuarine Macrobenthic.*

Dean, W., 1974. Determination of carbonate and organic matter in calcareous sediments and sedimentary rocks by loss on ignition: comparison with other methods. *Journal Sediment Petrology* 44, 242–248.

Di Lorenzo, E., Miller, A., Schneider, N. Mc Williams, J. 2005. The warming of the California Current system – Dynamics and ecosystem implications. *J. Physical Oceanography* 35: 336-362.

Diaz del Castillo, B. 1568. Historia verdadera de la Conquista de la Nueva España. Available at: http://www.cervantesvirtual.com/servlet/SirveObras/ 01715418982365098550035 /index.htm

Diaz, R., M. Solan and R. Valente. 2004. A review of approaches for classifying benthic habitats and evaluating habitat quality. *J. Environm. Management* 73:165-181.

Díaz-Castañeda, V. and V. Rodríguez. 1998. Polychaete fauna from San Quintín Bay, Baja California, Mexico. *Bull. Southern California Acad. Sci. Vol.* 97: 9-32.

Díaz-Castañeda, V. and D. Reish. 2008. Polychaetes in Environmental Studies. In: *Annelids as Models Systems in the Biological Sciences.* Ed. D. Shain. John Wiley and Sons. In press.

Durham, J.W. 1947. Corals from the Gulf of California and the north Pacific coast of America. *Geological Society of America Memoir* 20: 1-68.

Durham, J.W. y J.L. Barnard. 1952. Stony corals of the eastern Pacific collected by the Velero III and Velero IV. Allan Hancock Pacific Expeditions 16: 1-110.

Ebeling, A. W. (1967). Zoogeography of tropical deep sea animals. *Studies in Tropical Oceanography*, 5, 593-613.

Edwards, M. 2004. Estimating scale dependency in disturbance impacts: El Niños and giant kelp forests in the northeast Pacific. *Oecologia* 138: 436-447.

Escribano, R., G. Daneri, L. Farías, V. Gallardo, H. González, D. Gutiérrez, C. Lange, C. Morales, O. Pizarro, O. Ulloa and M. Brauni. 2004. Biological and chemical consequences of the 1997–1998 El Niño in the Chilean coastal upwelling system: a synthesis. *Deep-Sea Research II* 51: 2389–2411.

Färber-Lorda J., A. Trasviña, P. Cortes. (Sumited) Summer Distribution of Euphausiids in the Entrance of the Sea of Cortés in relation to hydrography. *Deep-Sea Research II.*

Färber-Lorda, J., Lavín, M. F.and Guerrero, M. A. (2004b) Effects of wind forcing on the trophic conditions, zooplankton biomass and krill biochemical composition in the Gulf of Tehuantepec . *Deep-Sea Research II*, 51, 615-627.

Färber-Lorda, J., Lavin, M. F., Zapatero, M. A., and Robles, J. M. (1994). Distribution and abundance of euphausiids in the Gulf of Tehuantepec during wind forcing. *Deep-Sea Research*, 38, 359-367.

Färber-Lorda, J., Trasviña, A. and Cortés-Verdín, P. (2004c) Trophic conditions and zooplankton distribution in the entrance of the Sea of Cortés, during summer. *Deep-Sea Research II*, 51, 601-614.

Farrel TM, Bracher D, Roughgarden J .1991. Cross-shelf transport causes recruitment to intertidal populations in central California. *Limnology and Oceanography* 36: 279-288.

Farreras, S. 2006. Hidrodinámica de Lagunas Costeras. *CICESE.* 184 pp.

Fernández-Alamo, M., Färber-Lorda J., (2006) Zooplankton and the Oceanography of the Eastern Tropical Pacific: A review. Progress in Oceanography, 69, 318-359.

Fernández-Álamo, M. A., Sanvicente-Añorve, L., and Alameda-de-la-Mora, G. (2000) Copepod assemblages in teh Gulf of Tehuantepec, México. *Crustaceana*, 73, 9,1139-1153.

Fernández-Alamo, M. 1987. Distribuciòn y abundancia de de los poliquetos pelàgicos (Annelida-Polychaeta) en el Golfo de Tehuantepec, Mèxico. In: Gòmez –Aguirre, S., Arenas-Fuentes, V. (Eds.), Contribuciones en Hidrobiología. Universidad Nacional Autònoma de Mèxico, Mèxico, pp. 269-278

Fields, P. A., J. B. Graham, R. H. Rosenblatt, and G. N. Somero. 1993. Effects of expected global climate change on marine faunas. Trends in Ecology and Evolution 8: 361-367.

Franco-Gordo, C. (2001). *Asociaciones de larvas de peces en aguas costeras del Pacífico Central Mexicano.* Ph D Thesis. Universidad de Colima, México. 175 pp.

Franco-Gordo, C., Flores-Vargas, R., Navarro-Rodríguez, C., Funes-Rodríguez, R., and Saldierna-Martínez, R. (1999). Ictioplancton de las costas de Jalisco y Colima, México (diciembre de 1995 a diciembre de 1996). *Ciencias Marinas*, 25, 107-118.

Forsbergh, E. D., and Reid, J. (1964) Biological production in the eastern Pacific Ocean. *Bulletin Inter-American Tropical Tuna Comission* 8, 479-527.

Franco-Gordo, C., Godìnez-Domìnguez, E., Filonov, A. E., Tereshchenko, I. E., Freire, J., (2004) Plankton biomasa and larval fish abundante prior to and during the El Niño period of 1997-1998 along the central Pacific coast of Mèxico. *Progress in Oceanography*, 63, 99-123.

Gaines S, Brown S, Roughgarden J .1985. Spatial variation in larval concentrations as a cause of spatial variation in settlement for the barnacle, Balanus glandula. *Oecologia* 67: 267-272.

Gallardo, V., M. Palma, F. carrasco, D. Gutiérrez, L. Levin and J. Cañete. 2004. Macrobenthic zonation caused by the oxygen minimum zone on the shelf and slope off central Chile. *Deep-Sea Res*. II 51: 2475-2490.

Glynn, P.W. y J.S. Ault. 2000. A biogeographic analysis and review of the far eastern Pacific coral reef region. *Coral Reefs* 19: 1-23.

Gonzalez-Silvera, A., Santamaría-del-Angel, E., Millan-Nuñez, R., Manzo-Monroy, H., 2004. Satellite observations of mesoscale eddies in the Gulfs of Tehuantepec and Papagayo (eastern Tropical Pacific). *Deep-Sea Research II,* 51, 587-600.

Hastings, P. A. (2000). Biogeography of the tropical eastern Pacific: distribution and phylogeny of chaenopsid fishes. *Zoological Journal of the Linnean Society*, 128, 319-335.

Helly, J. J.,and Levin, L. (2004) Global distribution of naturally occurring marine hypoxia on continental margins. *Deep-Sea Research I*, 51, 1159-1168.

Hernandez-Carmona, G., Robledo, D., and Serviere-Zaragoza, E. 2001. Effect of nutrient availability on Macrocystis pyrifera recruitment and survival near its southern limit off Baja California. *Botanica Marina* 44 : 221-229.

Hobday, A., and Tegner, M. J., 2000. Status review of white abalone .Haliotis sorenseni. throughout its range in California and Mexico. NOAA Technical Memorandum NOAA-TM-NMFS-SWR-035. U.S. Department of Commerce.

Holmes, R. W. (1958) Physical, chemical and biological oceanographic observations obtained on Expedition Scope in the eastern tropical Pacific, November-December 1956. *U. S. Fish and Wildlife Service, Special Scientific Report Fisheries*, Vol. 279, 117pp.

Ibarra-Obando, S., S. V. Smith, M. Poumian-Tapia, V. Camacho-Ibar, J. D. Carriquiry and M. Montes-Hugo. 2004. Benthic metabolism in San Quintin Bay, Baja California, Mexico. *Mar Ecol Prog Ser.* Vol. 283: 99-112.

Johnson, M. W., and Brinton, E. (1963). Biological species, water-masses, and currents. In M.N: Hill, *The Sea*. (381-414). New York: Wiley.

Kamykowski, D. and Zentara, S. 1990. Hypoxia in the world ocean as recorded in the historical data set. *Deep-Sea Research* 37: 1861–1874.

Kamykowski, D., and Zentara S-J. (1990) Hypoxia in the world ocean as recorded in the historical data set. *Deep-Sea Research*, 37, 12, 1861-1874.

Kessler, W. S. (2004). The circulation of the eastern tropical Pacific: A review. *Progress in Oceanography*,.in prep.

Ketchum, J.T. y H. Reyes Bonilla. 2001. Taxonomía y distribución de los corales hermatípicos (Scleractinia) del Archipiélago de Revillagigedo, Pacífico de México. *Revista de Biología Tropical* 49: 803-848.

Ladah, L. and Leichter, J., in review. High-frequency vertical thermocline motions as a nutrient source for kelp at its southern limit in the eastern North Pacific.

Ladah, L. and Zertuche, J., in review. Comparison of El Nino events in the past decade in giant kelp forests in Mexico.

Ladah, L., 2003. The shoaling of nutrient-enriched subsurface water as a mechanism to sustain primary productivity off Central Baja California during El Niño winters. *J. Mar. Systems*, 42: 145-152.

Ladah, L., Zertuche-Gonzalez, J. and Hernandez-Carmona, G., 1999. Giant kelp .Macrocystis pyrifera, Phaeophyceae. recruitment near its Southern Limit in Baja California after mass disappearance during ENSO 1997-1998. *J. Phycol.*, 35:1106-1112.

Ladah, L.and Tapia, F., in review. Small-scale temporal and spatial variability variability of Chthamalus spp. settlement and its connection with high-frequency physical phenomena in Northern Baja California.

Ladah, L.B., in review. Population structure of intertidal Egregia .Turner. on a wave-exposed shore in Northern Baja California during El Niño 2002-2004.

Ladah, L.B., and J.A. Zertuche. 2004. Giant kelp .Macrocystis pyrifera. survival in deep water .25-40m. during El Niño of 1997-1998 in Baja California, México. Botanica Marina, 47:367-372.

Ladah, L.B., F. J. Tapia, J. Pineda, and M. López, 2005. Spatially heterogeneous, synchronous settlement of Chthamalus spp. larvae in Northern Baja California. *Marine Ecology Progress Series* 302: 177-185.

Landry, M. R., Barber, R. T., Bidigare, R. R., Chai, F., Coale, K. H., Dam, H. G., Lewis, M. R., McCarthy, J. J., Roman, M. R., Stoecker, D. K., Verity, P. G., and White, J. R., (1997) Iron and grazing constrains on primary production in the central equatorial Pacific. *Limnology and Oceanography*, 42, 405-418.

Landry, M. R., Ondrusek, M. E., Tanner, S. J., Brown, S. L., Constantinou, J., Bidigare, R. R., Coale, K. H., Fitzwater, S., (2000a) Biological response to iron fertilization in the eastern equatorial Pacific (Ironex II). I.- Microplankton community abundances and biomass. *Marine Ecology Progress Series*, 201, 27-42.

Landry, M. R., Constantinou, J., Latasa, M., Brown, S. L. ,Bidigare, R. R., and Ondrusek, M. E. (2000b) Biological response to iron fertilizationin the eastern equatorial Pacific (IronEx II). III.- Dynamics of phytoplankton growth and microzooplankton grazing. Marine Ecology Progress Series 201, 57-72.

Lavin, M., P. Fiedler, J.A. Amador, L. Ballance, J. Farber and A. Mestas. 2006. A review of eastern tropical Pacific oceanography: Summary. Progress in Oceanography 69: 391-398.

Le Fèvre J, Bourget E .1992. Hydrodynamics and behaviour: transport processes in marine invertebrate larvae. *Trends in Ecology and Evolution* 7: 288-289

Leichter, J.J., Stewart, H., and Miller, S.L., 2003. Episodic nutrient transport to Florida coral reefs. *Limnol. Oceanogr.* 48: 1394-1407.

Leppaekoski, E. 1975. Assessment of degree of pollution on the basis of macrozoobenthos in marine and brakish-water environments. *Acta Acad. Aabo*, Vol. 35: 1-96.

Levin, L. and J. Gage. 1998. Relationships between oxygen, organic matter and the diversity of bathyal macrofauna. *Deep-Sea Research* 45, 129–163.

Levin, L. 2003. Oxygen minimum zone benthos: adaptation and community response to hypoxia. *Oceanogr. Mar. Biol. Ann. Rev.*, 41: 1–45.

Love, C. M. (ed.) (1972-1978) EASTROPAC ATLAS. National Ocean and Atmosphere Administration. *NFSW.*, Vols. 1 to 11.

Macho G, Molares J, Vazquez E .2005. Timing of larval release by three barnacles from the NW Iberian Peninsula. *Marine Ecology Progress Series* 298: 251-260

Mann, R., 1984. On the selection of aquaculture species: a. case study of marine molluscs. *Aquaculture* 39: 345-353.

McGowan, J. A. (1960a). The Systematic, Distribution, and Abundance of Euthecosomata of North Pacific. PhD Thesis, University of California, San Diego. 1-212.

McGowan, J. A. (1960b). The relationship of the distribution of the pelagic worm, *Poeobious meseres* Heath, to the warmer masses of the North Pacific. *Deep-Sea Research*, 6, 125-139.

McGowan, J. A. (1967). Distributional atlas of pelagic molluscs in the California Current region. *California Cooperative Oceanic Fisheries Investigations, Atlas* 6, 218 pp.

McGowan, J. A. (1971). Oceanic Biogeography of the Pacific. In B.M. Funnell and W.R. Riedel. *Micropaleontology of Oceans*. (pp 3-74). London: Cambridge University Press.

McGowan, J. A. (1974). The nature of oceanic ecosystems. In C.B. Miller, *The Biology of the Oceanic Pacific*. (pp 9-28). Corvallis, Oregon State University Press.

McGowan, J. A. (1986). The Biogeography of pelagic ecosystems. In A.C. Pierrot-Bults, S. van der Spoel, B. J. Zahuranec and R.K. Johnson, *Pelagic Biogeography*. UNESCO *Technical Papers in Marine Science* 49: 191-200.

McNeely, J., K. Miller, W. V. Reid, R. A. Mittermeier, T. B. Werner. 1990. Conserving the World's Biological Diversity, p. 193, IUCN, Gland, Switzerland.

McPhee-Shaw, E., Siegel, D., Washburn, L., Brzezinski, M., Jones, J., Leydecker, A., Melack,. J. 2007. Mechanisms for nutrient delivery to the inner shelf: observations from the Santa Barbara Channel. *Limnology and Oceanography* 52.5.: 1748-1756.

Mermillod-Blondin F. and R. Rosenberg. 2006. Ecosystem engineering: the impact of bioturbation on biogeochemical processes in marine and freshwater benthic habitats. *Aquat. Sci.* Vol. 68: 434–442.

Norris KS .1963. The function of temperature in the ecology of the percoid fish Girella nigricans .Ayres.. *Ecological Monographs* 33: 23-62.

Okubo, A., 1971. Oceanic diffusion diagrams. *Deep-Sea Res.*, 18: 789–802.

Olive, P. 1994. Polychaete as a world resource: a review of pattern of explotation as sea angling baits and the potential for aquaculture based production. In: Actes 4ème Conférence International Polychètes. Memoires Muséum National d'Histoire Naturelle. J. C. Dauvin, L. Laubier and D. Reish (eds.) 162: 603-610. MNHN, Paris.

Pares Sierra A, López M, Pavía E. 1997. Oceanografía Física del Océano Pacífico Nororiental. In: Lavín MF(ed) Contribuciones a la Oceanografía Física en México. Unión Geofísica Mexicana, pp 1-24.

Pearson, T. H. and R. Rosenberg. 1987. Feast and famine: structuring factors in marine benthic communities: 373-395. In: Gee J. and P. Giller (eds). *Organization of communities: past and present. The 27th Symposium of the British Ecological Society.* Blackwell Scientific Publications. Oxford.

Pennington, J. T., Mahoney, K. L., Kuwahara, V. S., Kolber, D. D., Calines, R., and Chavez, F., (2006) Primary production in the eastern tropical Pacific: A review. *Progress in Oceanography*, 69, 285-317.

Peterson, B. 1999. Stable isotopes as tracers of organic matter input and transfer in benthic food webs: A review. *Acta Oecologica* 20: 479-487.

Pineda J .1991. Predictable upwelling and the shoreward transport of planktonic larvae by internal tidal bores. *Science* 253: 548-551

Pineda J .1994a. Internal tidal bores in the nearshore: warm-water fronts, seaward gravity currents and the onshore transport of neustonic larvae. *Journal of Marine Research* 52: 427-458

Pineda J .1994b. Spatial and temporal patterns in barnacle settlement rate along a southern California rocky shore. *Marine Ecology Progress Series* 107: 125-138.

Pineda J .1995. An internal tidal bore regime at nearshore stations along western U.S.A.: predictable upwelling within the lunar cycle. Continental Shelf Research 15: 1023-1041.

Pineda J .1999. Circulation and larval distribution in internal tidal bore warm fronts. *Limnology and Oceanography* 44: 1400-1414.

Pineda J .2000. Linking larval settlement to larval transport: assumptions, potentials, and pitfalls. In: Färber-Lorda J .ed. *Oceanography of the Eastern Pacific*. CICESE, Ensenada, pp 84-105.

Pocklington, P. and P. Wells. 1992. Polychaetes:Key taxa for marine environmental quality monitoring. *Mar. Pollut. Bull.* Vol. 24: 593-598.

Queiroga H, Blanton J .2005. Interactions between behaviour and physical forcing in the control of horizontal transport of decapod crustacean larvae. *Advances in Marine Biology* 47: 107-214.

Rainbow, P. 1995. Biomonitoring of heavy metal availability in the marine environment. *Mar. Poll. Bull.* 31: 183-192.

Reid, J. L. (1962). On circulation, phosphate-phosphorus content, and zooplankton volumes in the upper part of the Pacific Ocean. *Limnology and Oceanography*, 7, 287-306.

Reid, J. L., Brinton, E., Fleming, A., Venrick, E. L., and McGowan, J. A. (1978). Ocean Circulation and marine life. In H. Charnock, G. Deacon, *Advances in Oceanography* (pp 65-130). New York: Plenum.

Reyes Bonilla, H. 2001. Effects of the 1997-1998 El Niño-Southern Oscillation on coral communities of the Gulf of California, México. *Bulletin of Marine Science* 69: 251-266.

Reyes Bonilla, H. 2003. *Coral reefs of the Pacific coast of México*. pp 331-349, En: J. Cortés (ed.). Coral reefs of Latin America. Elsevier, Amsterdam.

Reyes Bonilla, H. y G. Cruz Piñón. 2000. Biogeografía de los corales ahermatípicos (Scleractinia) del Pacífico de México. *Ciencias Marinas* 26: 511-531.

Reyes Bonilla, H., L.E. Calderon Aguilera, G.Cruz Piñon, P. Medina Rosas, R.A. López Pérez, M.D. Herrero Pérezrul, G. E. Leyte Morales, A. L. Cupul Magaña, J.D. Carriquiry Beltrán. 2005. Atlas de los corales pétreos (Anthozoa: scleractinia) del Pacífico Mexicano. Centro de Investigación Científica y de Educación Superior de Ensenada, Comisión Nacional para el Conocimiento y Uso de la Biodiversidad, Consejo Nacional de Ciencia y Tecnología, Universidad de Guadalajara / Centro Universitario de la Costa, Universidad del Mar. ISBN970-27-0779. 128 pp.

Rhoads, D. and J. Germano. 1986. Interpreting long-term changes in benthic community structure: a new protocol. Hydrobiologia, 142: 291-308.

Rice, J. 2003. Environmental health indicators. Ocean and Coastal Management. 46: 235-259.

Robinson CJ, Gómez-Aguirre S, Gómez-Gutiérrez J .2007. Pacific sardine behaviour related to tidal current dynamics in Bahía Magdalena, Mexico. *J. Fish Biol* 71:200–218.

Rollwagen-Bollens, G. C., and Landry, M. R. (2000). Biological response to iron fertilization in the eastern equatorial Pacific (IronEx II). 2. Mesozooplankton abundance, biomass, depth distribution and grazing. *Marine Ecology Progress Series* 201, 43,56.

Ryggs, B. 1985. Distribution of species along pollution-induced diversity gradients in benthic communities in Norwegian Fjords. *Mar. Pollut. Bull.*, 16: 469-474.

Saavedra-Sotelo, N.C. 2007. Estructura y flujo genético de Pavona gigantea (Anthozoa: Scleractinia) en las costas del Pacífico mexicano. M. Sc Thesis, CICESE, Ensenada, 65 pp.

Saltzman, J., and Wishner, K. F. (1997a). Zooplankton ecology in the eastern tropical Pacific oxygen minimum zone above a seamount: 1.- General trends. *Deep-Sea Research I*, 44, 6, 907-930.

Sandstrom, H., and J. A. Elliott. 1984. Internal tide and solitons on the Scotian shelf: a nutrient pump at work. *Journal of Geophysical Research* 89: 6415-6426.

Sanford, E. and B.A. Menge. 2007. Reproductive output and consistency of source populations in the sea star Pisaster ochraceus. *Marine Ecology Progress Series* 349: 1–12.

Segura-Puertas, L. (1987). Distribución y abundancia de las Medusas (Cnidaria: Hydrozoa y Scyphozoa) en el Golfo de Tehuantepec, México. En S. Gómez-Aguirre y V. Arenas-Fuentes, *Contribuciones en Hidrobiología.* (pp 260-267). México: Universidad Nacional Autónoma de México.

Semenov, V. N., and Berman, S. (1977) Biogeographic aspects of the distribution and dynamics of water masses off the South American coasts. *Oceanology* 17, 1073-1084.

Shanks AL .1983. Surface slicks associated with tidally forced internal waves may transport pelagic larvae of benthic invertebrates and fishes shoreward. Marine Ecology Progress Series 13: 311-315.

Shanks AL, Largier JL, Brink L, Brubaker J, Hoof R. 2000. Demonstration of the onshore transport of larval invertebrates by the shoreward movement of an upwelling front. *Limnology and Oceanography* 45: 230-236.

Shanks AL. 1995. Mechanisms of cross-shelf dispersal of larval invertebrates and fish. In: McEdward L .ed. *Ecology of Marine Invertebrate Larvae*. CRC Press, Boca Raton, FL, pp 323-367.

Smith, K. 1989. Short time series measurements of particulate organic carbon flux and sediment community oxygen consumption in the North Pacific. *Deep-Sea Res.*, 36, 1111-1119.

Squires, D.F. 1959. Corals and coral reefs in the Gulf of California. *Bulletin of the American Museum of Natural History* 118: 367-432.

Steuer, A. 1933. Zur planmässingen Erfoschung der geographishen Verbreitung des Haliplanktons, besonders Copepoden. *Zoogeographica International Review Comparative causal Animal Geographie*, 1, 269-302.

Sverdrup, H. V., Johnson, M. W., andFleming, R. H. 1942. *The oceans: their physics, chemistry, and general biology*. New York: Prentice-Hall. 1087 pp.

Tont, S. 1976. Short period climatic fluctuations effects on diatom biomass. *Science* 194:942-944.

Trasviña A, Andrade CA. 2002. La circulación costera del Pacifico Tropical Oriental, con énfasis en la Alberca Cálida Mexicana (ACM). In: *Circulación oceánica y climatología tropical en México y Colombia*, Corcas Editores, Bogotá pp 9 -37.

Tsutsumi, H. 1990. Population persistence of *Capitella* sp. (Polychaeta; Capitellidae) on a mud flat subject to environmental disturbance by organic enrichment. *Marine Ecology Progress Series* 63: 147-156.

Van der Spoel, S., and Heyman, R. P. 1983. *A comparative atlas of zooplankton, biological patterns in the oceans*. Berlin: Springer. 186 pp.

Van der Spoel, S., and Pierrot-Bults, A.C. 1979. *Zoogeography and diversity of plankton.* Utrech: Bunge. 410 pp.

Vermeij, G. J. 1978. Biogeography and adaptation: Patterns of marine life. Harvard University Press, Cambridge, 332 pp.

Veron, J.E.N. 2000. *Corals of the World.* Vols. 1-3. Australian Institute of Marine Science, Townsville.

Verrill, A.E. 1864. List of the polyps and corals sent by the Museum of Comparative Zoology to other institutions in exchange, with annotations. *Bulletin of Museum Comparative Zoology*, Harvard 1: 29-60.

Verrill, A.E. 1868. Review of the corals and polyps of the west coast of America. Transactions of the Connecticut Academy of Arts and Sciences 1:377-558.

Villalobos, A. 1960. Notas acerca del aspecto hidrobiológico de la isla. pp. 154-180. En: J. Adem, E. Cobo, L. Blásquez, A. Villalobos, E. Miranda, T. Herrera, B. Villa y L. Vázquez (eds.). La Isla Socorro, Archipiélago de Revillagigedo. Monografías del Instituto de Geofísica, U.N.A.M. 2.

Wieking, G. and I. Kröncke. 2005. Is benthic trophic structure affected by food quality? The Dogger Bank example. *Mar. Biol.*, 146: 387-400.

Willett, C. S., Leben, R., Lavin, M. F., 2006. Eddies and tropical instability waves in the eastern tropical Pacific: A review. *Progress in Oceanography*, 69, 218-238.

Wing SE, Largier JL, Botsford LW, Quinn JF .1995a. Settlement and transport of benthic invertebrates in an intermittent upwelling region. *Limnology and Oceanography* 40: 316-329.

Wing SR, Botsford LW, Largier JL, Morgan LE. 1995b. Spatial structure of relaxation events and crab settlement in the northern California upwelling system. *Marine Ecology Progress Series* 128: 199-211.

Witman, J.D., Leichter, J.J., Genovese, S.J., and Brooks, D.A. 1993. Pulsed phytoplankton supply to the rocky subtidal zone: influence of internal waves. *Proc. Natl Acac. Sci. USA*. 90: 1686-1690.

Zaytsev O, R. Cervantes-Duarte, O. Sánchez-Montante, A. Gallegos-García. 2003. Coastal upwelling activity on the Pacific shelf of the Baja California peninsula. *J Oceanog.* 59:489–502.

Zimmerman, R. and Robertson, D. 1985. Effects of El Niño on local hydrography and growth of giant kelp Macrocystis pyrifera, at Santa Catalina Island, California. *Limnology and Oceanography* 30: 1298–1302.

Zimmerman, R.C. and Kremer, N.J., 1984. Episodic nutrient supply to a kelp forest ecosystem in Southern California. *J. Mar. Res.*, 42:591-604.

In: The Pacific and Arctic Oceans
Editor: Kallen B. Tewles

ISBN: 978-1-60692-010-7

Chapter 2

TEMPORAL AND SPATIAL VARIATIONS OF VERTICAL DISTRIBUTIONS OF NUTRIENTS IN THE KUROSHIO RECIRCULATION REGION OF THE WESTERN NORTH PACIFIC

K. Hirose*

Geochemical Research Department, Meteorological Research Institute,
Nagamine 1-1, Tsukuba, Ibaraki 305-0052 Japan

ABSTRACT

Temporal and spatial variations of vertical profiles of nitrate and phosphate concentrations in seawater in the Kuroshio recirculation (KR) region water columns were analyzed by using a biogeochemical curve fitting method, which includes a function constructed from constant upwelling velocity, and simple biogeochemical processes with the first order kinetics of nutrient consumption and regeneration and particle export flux based on an assumption of steady state. The simple biogeochemical equation was applied to vertical profiles of nutrient concentrations for WOCE P3 (130°-160°E) and Station P9-B (25°N, 137°E) time series data from 1981 to 2002. Most of the vertical profiles of the nitrate and phosphate concentrations can fit well with the biogeochemical equation with correlation factors of >0.998. The spatial distribution of obtained upwelling velocities corresponds to that of meso-scale eddies in the KR region. The analysis of time series data using the curve fitting method provides evidence for inter-annual variations of nutrient profiles as does N:P ratio, which is primarily led by the variability of physical processes such as the variations of generation of meso-scale eddies and its southwest ward motion of meso-scale eddies. The variation of nutrients may be associated with the change of phytoplankton community structure following marine ecological change.

Key words: nutrient, nitrate, phosphate, vertical profile, N:P ratio, interannual variability

---

* Corresponding author. Tel: +81-29-853-8718; Fax: +81-29-853-8728. Email address: khirose@mri-jma.go.jp (K. Hirose)

INTRODUCTION

Biogeochemical process, such as biological pumping, directly related to carbon cycling in the ocean is one of controversial topics in the field of global environmental change. Nutrients in seawater are directly related to oceanic biogeochemical processes because nutrients are closely related to primary production by phytoplankton and particle export flux to the ocean interior (Broecker and Peng, 1982). Especially, to have better understanding of oceanic behaviors of nutrients, i.e., nitrate and phosphate, is one of the most important issues because nitrate and phosphate in seawater are considered to be a limiting factor on biological production in the major part of open ocean areas (Broecker and Peng, 1982; Tyrrell, 1999).

Temporal variations of nitrate and phosphate concentrations in the ocean are one of the current topics related to the ecological effects of climate change to the global environment. Regarding variation of nutrients in the ocean, we have two important problems: one, is that N:P ratio in deep waters should be equal to the canonical Redfield ratio (N:P =16)(Redfield et al., 1963), and another, is whether N:P ratio of dissolved inorganic nutrients shows temporal variation or not, because the constant N:P stoichiometry has been used in biogeochemical modeling (Van Cappellen and Ingall, 1994). The N:P ratio in the deep ocean in all three ocean basins is deserved to be 14.7, which is significantly lower than the Redfield N:P ratio. The difference has been explained by denitrification in the deep ocean (Gruber and Sarmiento, 1997) and change of nitrogen fixation by phytoplankton over time (Falkowski, 1997; Karl et al., 1997). As an alternative explanation, the deep N:P ratio is determined by remineralization of mixture of biological species with lower N:P ratios (typically, diatoms) and species with higher N:P ratios (Rubin, 2003). Although spatial and vertical variations of N:P ratios have been recognized (Watson and Whitfield, 1985; Fanning, 1992), temporal differences in deep nutrient measurements between subsequent cruises were generally considered to be analytical errors (Broecker et al., 1985; Anderson and Sarmiento, 1994; Gruber and Sarmiento, 1997). However, Pahlow and Riebesell (2000) suggested that the deep water N:P ratios in the northern hemisphere show evidence for temporal trends over the past five decades. Ecological change in the ocean leads to the variation of the N:P ratio in deep waters. Therefore, it is significant to confirm whether nutrients in the water column vary temporarily or not. If the variation deduced from observation is natural variability, it is necessary to identify the physical and biogeochemical processes to cause the variability.

There is little information on the temporal variations of vertical profiles of nutrients associated with oceanographic changes such as regime shift (Mantua, 2004). The physical dynamics in the ocean such as mesoscale eddies and Rosby waves (Sakamoto et al., 2004) affect the vertical profiles of nutrients. According to the theoretical study of Flierl (1981), meso-scale eddies with a strongly nonlinear nature can trap a substantial amount of water parcels within them (greater than 30-40% of volume). Especially, mesoscale physical phenomena have a strong impact on the oceanic ecological structure and functioning. For example, path-through of cyclonic and anticyclonic eddies, in which upwelling and downwelling are occurred, respectively, leads to the change of nutrient concentrations in the water column. McGillicuddy et al. (1998) demonstrated the importance of intermittent upwelling events associated with eddies in supplying nutrients to the subtropical Atlantic near Bermuda, which has been recognized as eddy pumping (Jenkins, 1988; Falkowski et al., 1991; Garcçn et al., 2001).

The Kuroshio recirculation (KR) region is characterized by the presence of mesoscale eddies, which are generated in the vicinity of the Kuroshio extension region and move southwestward (Ebuchi and Hanawa, 2000, 2001). Climate change may affect generation and transport of mesoscale eddies in the Kuroshio extension region of the western North Pacific. Therefore, it is likely that a significant oceanic variability of nutrients in the water column occurs in the KR region of the western North Pacific as a result of climate change. The vertical profiles of nutrient concentrations in the KR region can be reproduced with high performance by a simple biogeochemical curve fitting function (Hirose and Kamiya, 2003); the curve fitting technique using the biogeochemical function can provide the physical and biogeochemical parameters, such as upwelling velocity, nutrient consumption rate and regeneration rate, and particle export flux from observed vertical profiles of nitrate and phosphate. These parameters are prone to solve physical and biogeochemical factors causing the variability of the vertical profiles of nutrients in the ocean.

In this paper, we describe the results of analysis of nutrients data based on a curve fitting method with the simple biogeochemical function, discuss the temporal spatial variations of vertical nutrient profiles and the N:P ratio, and their controlling biogeochemical and physical processes.

DERIVATION OF BIOGEOCHEMICAL CURVE FITTING FUNCTION

The differential equation governing the vertical distribution of the nutrient concentration $N_i(z)$ in the steady state is expressed as follows:

$$d/dz(wN_i(z)) - S_i(z) = 0 \quad (1)$$

where z denotes depth and $S_i(z)$ is a biogeochemical term reflecting sources and sinks of nutrients. $S_i(z)$ is constructed from the consumption of the nutrients, $P_i(z)$, and the regeneration of nutrients (or remineralization), $Q_i(z)$; $S_i(z) = Q_i(z) - P_i(z)$. The upwelling velocity, w, is assumed to be constant. The process of nutrient consumption in the water column is directly related to the primary production. We adopt a first-order kinetics for the consumption of nutrients to simplify the biological process; a similar simplification was used in the biogeochemical general circulation model (Yamanaka and Tajika, 1996), even though the Michaelis-Menten type formula has been usually used to describe uptake processes in the biological model (Franks et al., 1986; Maier-Reimer, 1993; Kishi, 1994). The differential consumption rate is given as

$$P_i(z) = d/dz\,(k_p(z)N_i(z)) \quad (2)$$

The depth-dependent consumption rate of nutrients due to the biological process, $k_{p(z)}$, is represented as an exponential function of depth (Franks et al., 1986; Kishi, 1994).

$$k_p(z) = k_o/\lambda_p \exp(-\lambda_p z) \quad (3)$$

where k_o (unit: yr^{-1})and λp (unit: m^{-1}) are respectively the consumption rate constant in the surface layer (100 m depth), which means a net consumption rate of nutrients supported by upwelling, and the depth-decrease constant for the consumption of nutrients due to biological processes.

The regeneration of nutrients in the water column, from the biological decomposition of sinking particles, is related to the export flux of particulate matter, Fi(z). We assume that the change of the export flux at depth corresponds to the input of nutrients remineralized in the water column as a result of decomposition by bacterial activity. The differential regeneration of nutrients in the water column is given as follows:

$$Q_i(z) = d/dz\, F_i(z). \tag{4}$$

The particle export flux is represented as an exponential function of depth (Shaffer, 1989):.

$$F_i(z) = F_{io} \exp(- \lambda_f z) \tag{5}$$

where F_{io} (unit: mol $m^{-2}yr^{-1}$) and λ_f (unit: m^{-1}) denote the export flux of N and P in the surface and the particle flux decay constant, respectively.

Substituting eq. 2-5 in eq. 1, we obtain a following differential equation:

$$d/dz(w(z)N_i(z)) + d/dz\, (k_o/\lambda_p \exp(-\lambda_p z)\, N_i(z)) - d/dz\, (F_{io} \exp(- \lambda_f z)) = 0 \tag{6}$$

This differential equation (eq. 6) can be solved under the conditions that concentrations of inorganic N and P in surface waters are negligible. Finally we have a biogeochemical curve fitting equation as a result of integration:

$$N_i(z) = A \exp\{(\lambda_p - \lambda_f)z \}/\{1 + B \exp(\lambda_p z)\} \tag{7}$$

where A and B are constant at each station; $A = F_{io}\lambda_p/k_o$ (unit: mol m^{-3}) and $B = w\, \lambda_p/k_o$ (non-dimensional parameter). Equation 7 has been applied to fit the vertical distribution of inorganic N and P in the western North Pacific using adjustable parameters, A, B, λ_p and λ_f.

DATA

Nutrient (nitrate and phosphate) concentrations in the water columns were obtained at Sta. P9-B (25°N 137°E) during the cruises of MS Ryofu-maru in the period 1981 to 2002, belonging to the Japan Meteorological Agency (JMA) (JMA, 1995; The Results of Marine Meteorological and Oceanographic Observations, No. 69 - 93). Nutrient data of World Ocean Circulation Experiment (WOCE) P3 stations (130°E – 160°E, along 24° 17'N) in 1985 was given by National Oceanic and Atmospheric Administration/Pacific Marine Environmental Laboratory (NOAA/PMEL). The sampling stations are shown in Figure 1. Nutrient analyses for JMA were carried out using a Technicon AutoAnalyzer^{TM}-II (AA-II). The analytical method is described in detail elsewhere (JMA, 1995; Kaneko et al., 1998; WOCE, 1994).

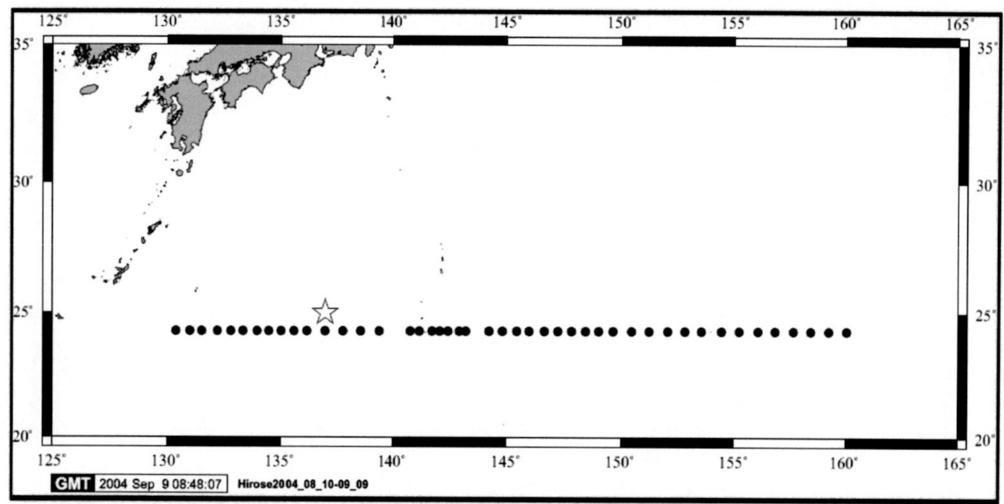

Figure 1. Sampling locations in the Kuroshio recirculation region of the western North Pacific. Star mark: Sta. P9-B, closed circle: WOCE P3.

RESULTS

Application of the Curve Fitting Method (the KR Regions):WOCE P3

The western cross section of nitrate concentrations along about 24° 17'N (WOCE P3) is shown in Figure 2. There are significant spatial variations of the nitrate and phosphate concentrations in water columns of the KR region, although a major pattern of the nitrate concentrations as does phosphate exhibits zonal distributions with a maximum layer of 1200 m depth. A major target is to have better understandings of factors controlling spatial variations of the nutrient concentrations in the KR region.

Individual vertical profile of nitrate and phosphate concentrations in the Kuroshio and KR region (19° – 34°N, along 137°E longitude) of the western North Pacific could be fitted with the biogeochemical function with high correlation factors (Hirose and Kamiya, 2003). This finding suggests that the vertical profiles of the nitrate and phosphate concentrations in the Kuroshio and KR region can be completely characterized by four parameters as did assumption of constant upwelling velocity be accepted. We applied the biogeochemical curve fitting method to WOCE P3 data (130°E – 160°E, along 24° 17'N latitude) in order to reconfirm applicability of the curve fitting function. Inorganic N (nitrate) and P (phosphate) concentrations from 100 m to around 2,500 m depth, including around 24 layers, were used for nonlinear curve fitting with the biogeochemical function for the KR region (45 stations from 130°E to 160°E). To apply the curve fitting function to observed vertical profiles of inorganic N and P, the particle flux decay constant, λ_f, is assumed to be independent of sampling locations because the degradation rate constant is related only to bacterial decomposition processes of sinking particles and bacterial densities remain relatively constant, irrespective of the productivity of the water column (Fenchel et al., 1998).

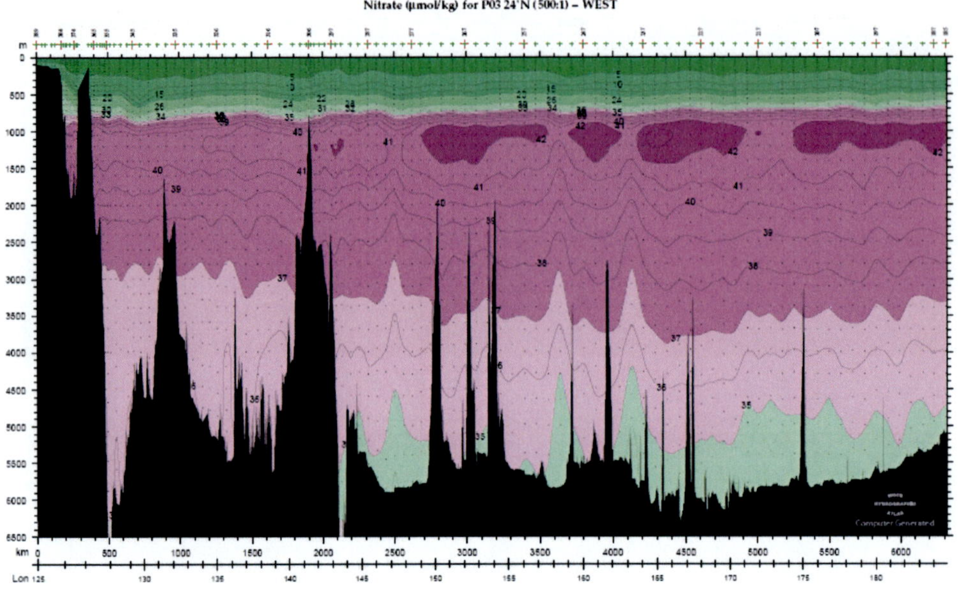

Figure 2. Western transect of nitrate concentrations along 24° 16'N (WOCE P3) latitude in the North Pacific (cited from WOCE Atlas Volume 2: Pacific Ocean.(WOCE, 2007)).

In order to confirm this assumption, the relationship between the particle flux decay constant and the correlation factor of the best-fit curve was examined and a constant value of 8×10^{-5} m^{-1} to an optimum value of the particle flux decay constant has been obtained in the western North Pacific (Hirose and Kamiya, 2003). Therefore we used this value as the particle flux decay constant, independent of the kind of nutrients and sampling stations in the KR region. The results of nonlinear least squares fitting for observed vertical profiles of nitrate and phosphate are shown in Figure 3. The biogeochemical equation (7) gives best-fit curves for the vertical profiles of nitrate and phosphate (45 stations along 24°17' N) with correlation factors greater than 0.998. Of 89 profiles fitted to the biogeochemical function, 91 % had correlation factors larger than 0.999. This result reconfirms that vertical profiles of nitrate and phosphate in the KR region of the western North Pacific can be very well reproduced by the simple biogeochemical function (eq. 7). The calculated results suggest that the depth-decrease constant for the consumption of inorganic N, λ_{pN}, had the same value as that of inorganic P (i.e., $\lambda_{pN} = \lambda_{pP}$). The λ_p (i.e., λ_{pN} or λ_{pP}) value, which is considered to depend on the biogeochemical conditions, such as phytoplankton distributions, varied from 0.61×10^{-2} to 0.75×10^{-2} m^{-1} in the KR regions.

Relationships between adjustable parameters obtained from the curve fitting technique showed that nitrate and phosphate exhibited relatively simple behavior in the ocean. We calculated the A/B ratios, which are equal to F_{io}/w. The A/B values for nitrate and phosphate ranged between 45.1 and 47.6 mmol m^{-3} with an average of 46.3±0.7 (N), and between 3.02 and 3.34 mmol m^{-3} with an average of 3.18±0.06 mmol m^{-3} (P), respectively. The F_{io}/w

values of nitrate in the WOCE P3 were the same range as that of WOCE P9, whereas the F_{io}/w values of phosphate were slightly lower than that of the WOCE P9.

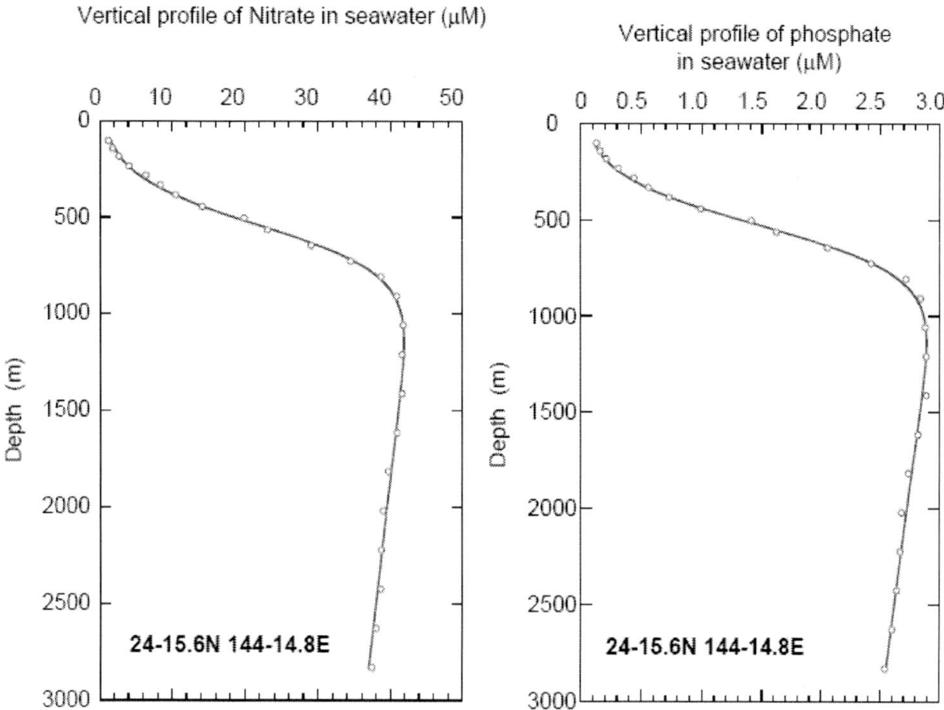

Figure 3. Vertical profiles of nitrate and phosphate at 24° 15.6'N 144° 14.8'E (WOCE P3). The solid curve is model fitting. The correlation factors of non-linear curve fitting for nitrate and phosphate were 0.9993 and 0.9995, respectively.

Time-Series Data at Sta. P9-B (25°N, 137°E)

In order to elucidate the temporal variation of vertical profiles of nutrients, we examined the simple model analysis to the time-series data of the KR station. The nitrate and phosphate concentrations from 100 m to around 2,500 m depth, including around 17 layers, were used for nonlinear curve fitting with the biogeochemical function for nutrient profiles in the KR region (94 profiles at Sta. P9-B 25°N, 137°E from 1981 to 2002). To apply the biogeochemical curve fitting function to observed vertical profiles of inorganic N and P, we used a constant value of 8×10^{-5} m^{-1} as an optimum value of the particle flux decay constant in the KR regions (Hirose and Kamiya, 2003).

The results of nonlinear least squares fitting for observed vertical profiles of the nitrate and phosphate concentrations are shown in Figure 4. The biogeochemical function gives best-fit curves for the vertical profiles of nitrate (94 profiles at Sta. P9-B 25°N, 137°E) with correlation factors greater than 0.998. Of 47 profiles of nitrate fitted to the biogeochemical function, 85 % had correlation factors larger than 0.999. On the other hand, 75 % of 47 profiles for phosphate was fitted to the biogeochemical function with correlation factors larger than 0.999 and 15 % had poor correlation factors less than 0.998. Occurrence of relatively poor correlation factors for phosphate may be attributable to analytical errors.

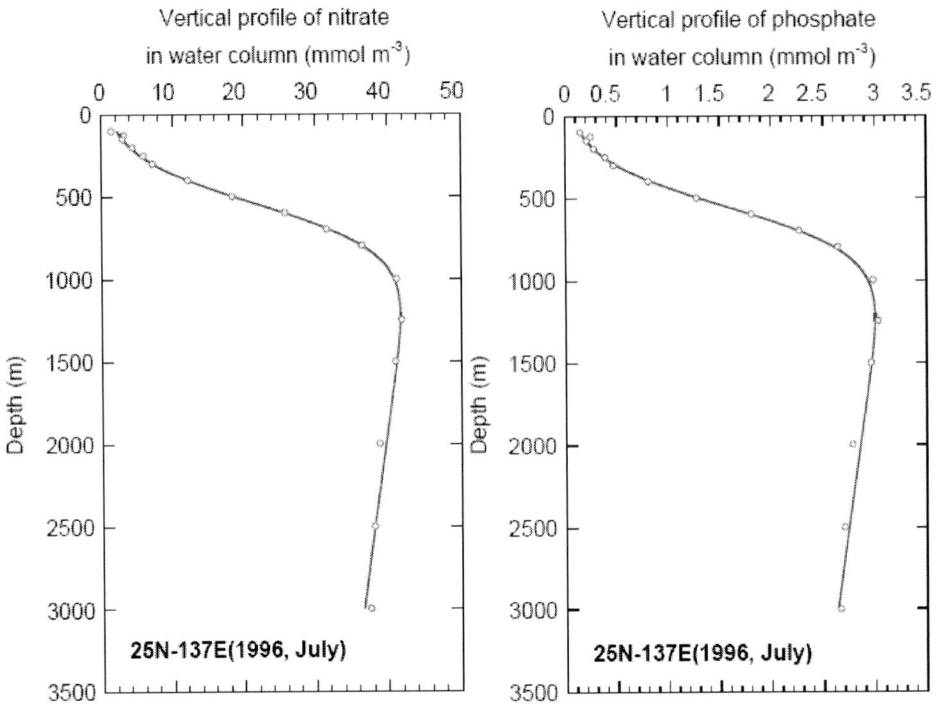

Figure 4. Vertical profiles of nitrate and phosphate at 25° 137° (WOCE P9-B) in July, 1996. The solid curve is model fitting. The correlation factors of non-linear curve fitting for nitrate and phosphate were 0.9996 and 0.9996, respectively.

As a result, past two decades observational data for vertical profiles of the nitrate and phosphate in the KR region of the western North Pacific can be very well described by the biogeochemical function (eq. 7). The calculated results confirm that the depth-decrease constant for the consumption of nitrate, λ_{pN}, had the same value as that of phosphate. The λ_p (i.e., λ_{pN} or λ_{pP}) values varied from 0.62×10^{-2} to 0.80×10^{-2} m^{-1} at Sta. P9-B, which were the same range as WOCE P9 and P3 in the KR regions.

Relationships between adjustable parameters obtained from the curve fitting give information on physical and ecological behaviors of nitrate and phosphate in the ocean. We calculated the A/B ratios, which are equal to F_{io}/w. The A/B values for nitrate and phosphate ranged between 45.2 and 48.5 mmol m^{-3} with an average of 46.7±0.6 (N), and between 3.14 and 3.53 mmol m^{-3} with an average of 3.32±0.09 mmol m^{-3} (P), respectively. The F_{io}/w values of nitrate and phosphate in the time-series data covered the spatial variation ranges for the WOCE P3 and P9.

Discussion

Physical Process Controlling Distribution of Nutrients: Upwelling Velocity

When the consumption rate of nutrients by marine organisms, k_O, is independent of oceanographic conditions, the upwelling velocity at each station can be calculated. The consumption rate of nutrients in our biogeochemical equation, corresponding to a net consumption rate of nutrients supported by upwelling, should not be identified with values determined by biological experiments (characterized by turnover times of several days). There is at present no information available on the value of the consumption rate constant of nutrients. Hirose and Kamiya (2003) introduced 1 yr^{-1} (0.0027 d^{-1}) as a typical consumption rate constant in the western North Pacific. We used this value to estimate upwelling velocities at WOCE P3 stations along 25°16'N latitude. The longitudinal distribution of upwelling velocities is shown in Figure 5. The upwelling velocities, ranging from 1.8 to 5.6 m y^{-1}, were in the same range as that in the KR region of WOCE P9 (Hirose and Kamiya, 2003), although upwelling velocities of WOCE P3 showed a larger spatial variation comparing with values of WOCE P9. The spatial width of peaks of the upwelling velocity was about 200 km, which is the same order of magnitude as that of the meso-scale eddies observed in the KR region (Ebuchi and Hanawa, 2002). The longitudinal variation of the upwelling velocities along WOCE P3 coincides with changes of water temperature and density; the low upwelling velocities correspond to thick warm water layers and thick low-density water layers, whereas the high upwelling velocity peak is consistent with a dome structure of water temperature, which implies cyclonic eddy.

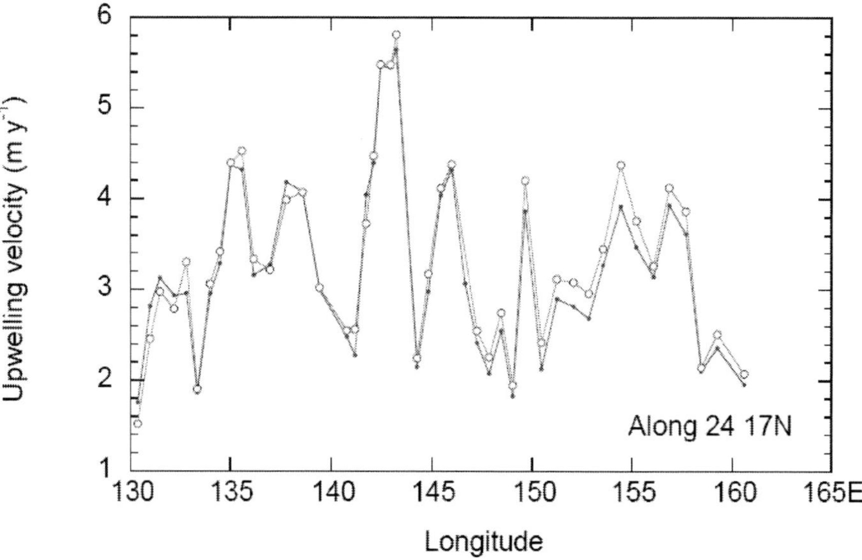

Figure 5. The longitudinal distribution of upwelling velocity along 24° 17'N (WOCE P3). Closed circle: values estimated from nitrate profiles assumption of k_{oN} = 1, open circle: values estimated from phosphate profiles assumption of k_{oP} = 1.

Especially the low upwelling velocities occurring near 130°E and 160°E corresponded to the regions with the thick low-nutrient layers as shown in Figure 2. These findings suggest that the spatial variation of the upwelling velocity reflects the spatial distribution of cyclonic and anticyclonic eddies. Therefore, the presence of the meso-scale eddies and their variability is attributable to causes to change the vertical profiles of the nutrient concentrations in the KR region of the western North Pacific.

Temporal Change of Upwelling Velocity

The upwelling velocities at Sta. P9-B in the period 1981 to 2002 were determined from the vertical profiles of nutrients using the biogeochemical curve fitting function and the consumption rate constant of 1 yr^{-1}. The temporal variation of upwelling velocity at Sta. P9-B is shown in Figure 6. The upwelling velocities at Sta. P9-B showed the interannual and decadal variability, whose amplitude was the same order of magnitude as observed in their longitudinal distributions (Figure 4). Peak of upwelling velocities may correspond to cyclonic eddies, whereas minima of upwelling velocities means anti-cyclonic eddies. In fact, Ebuchi and Hanawa (2001) revealed that about 5 anti-cyclonic and cyclonic eddies passed through a site (25°N, 135°E) during the five year period from Oct, 1992 to July 1998 based on TOPEX/POSEIDON altimeter data. The time-series data of upwelling velocity suggests that an inter-annual variation of upwelling velocity exists during the period of 1981 to 1992, when five cyclonic and anticyclonic eddies mutually passed through Sta. P9-B at about two year cycle. On the other hand, the inter-annual trend of upwelling velocity changed in the early 1990s.

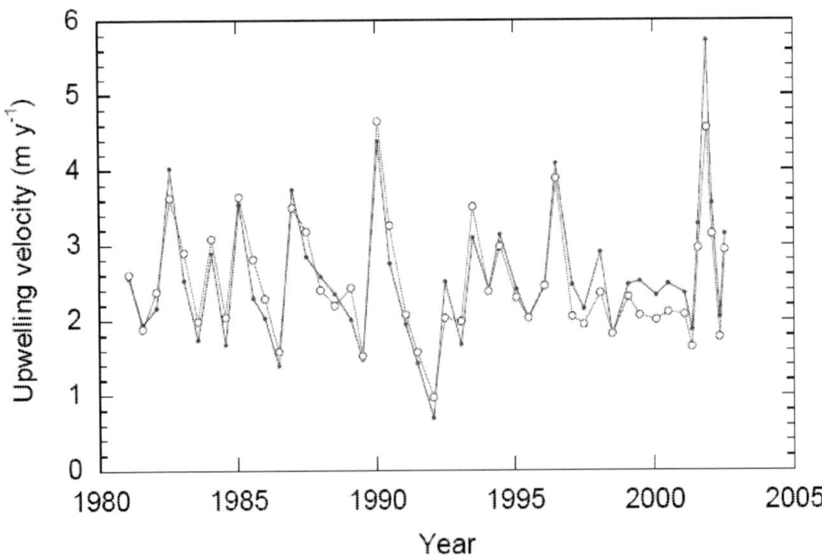

Figure 6. Temporal variation of upwelling velocity at Sta. P9-B during the period from 1981 to 2002. Closed circle: values estimated from nitrate profiles assumption of k_{oN} = 1, open circle: values estimated from phosphate profiles assumption of k_{oP} = 1.

Especially, there was no variation of upwelling velocity during the period of 1997 to 2001. This phenomenon may reflect the decadal change of meso-scale dynamics in the North Pacific. It, however, must be noted that the time series data of upwelling velocity cannot cover passage of all of meso-scale eddies because the observation was limited only two seasons (winter and summer from 1981 to 2000).

Biogeochemical Processes: Export Flux

Modeling studies (Oschlies and Garcçn, 1998; Garcçn et al., 2001) suggested that the presence of eddies generally contribute to enhance the biological productivity. According to application of the simple biogeochemical curve fitting technique to the vertical profile of nutrients, the export flux of nitrogen and phosphorus in the KR region can be estimated from calculated upwelling velocities and F/w values. Figure 7 shows the spatial distribution of the export flux of nitrogen (N export flux) from surface (100 m depth) to deep waters along the WOCE P3 (24° 17'N) line. The N export flux in the KR region ranged from 0.08 mol m^{-2} y^{-1} (0.2 mmol m^{-2} d^{-1}) to 0.26 mol m^{-2} y^{-1} (0.7 mmol m^{-2} d^{-1}). The N export flux showed larger spatial variability corresponding to that of the upwelling velocities. The result suggests that observation with higher spatial resolution such as WOCE observation is required in order to realize precise spatial distribution of the export flux into ocean interior. In case when a mean upwelling velocity in the North Pacific is about 3 m y^{-1} (Ganachaud and Wunsch, 2000), corresponding to a mean N export flux of about 0.14 mol m^{-2} y^{-1}, a N export flux driven by cyclonic eddies is estimated to be 0.18 mol m^{-2} y^{-1}, which is very similar to the eddy-induced upward flux in the Sargasso Sea (0.19 mol N m^{-2} y^{-1})(McGillicuddy et al., 1998). The N export flux in the KE region is enhanced approximately twice into cyclonic eddies.

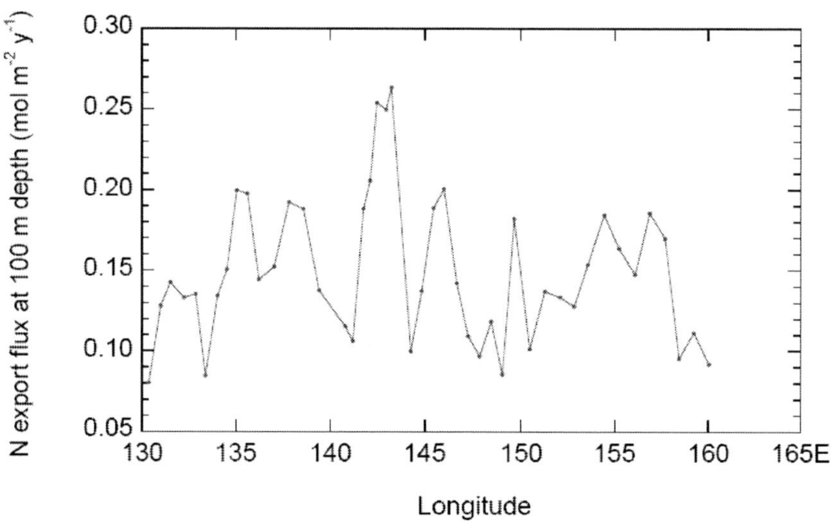

Figure 7. The longitudinal distribution of the N export flux along 24° 16'N (WOCE P3).

We examine the temporal variability of the N export flux in the KR region of the western North Pacific. The result at Sta. P9-B is shown in Figure 8. The N export flux at Sta. P9-B (range: 0.03 to 0.27 mol m^{-2} y^{-1}) exhibited interannual variability with the similar amplitude as observed in the spatial distribution in the KR region.

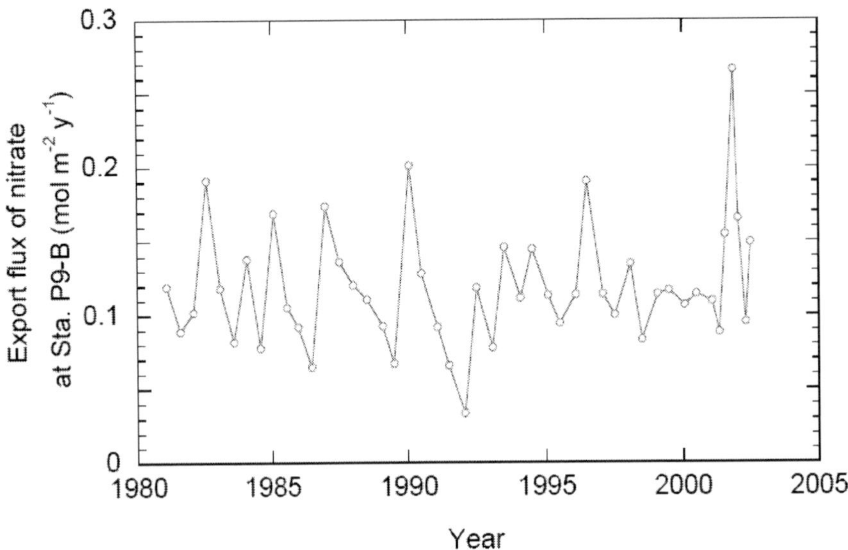

Figure 8. Temporal variation of the N export flux at Sta. P9-B during the period from 1981 to 2002.

This finding reveals that the N export flux in the KR region varied temporally according to the motion of cyclonic and anti-cyclonic eddies; especially, mesoscale biological activity is enhanced into cyclonic eddy. It must be noted that the amplitude of the variations of the N export flux in the KR region was the same order of magnitude as its temporal variation observed with the sediment trap experiment at Sta. ALOHA (Dore and Karl., 1996).

N:P Ratios of Dissolved Inorganic Nutirents

Redfield et al (1963) introduced C:N:P stoichiometry (106:16:1) to summarize the earlier research of dissolved and particulate nutrient pools in the ocean. The N:P ratios in inorganic nutrients in seawater have been considered to reflect N:P stoichiometry in marine organisms. However, the variability of the N:P ratios in real inorganic nutrient pools indicates that N:P stoichiometry is not simply formed by production or decomposition of marine organisms. In fact, laboratory studies revealed that the Redfield stoichiometry is achieved only during maximum growth rate in nutrient-sufficient cultures (Goldman at al., 1979; Laws and Bannister, 1980). The so-call Redfield ratio is used as an important unifying concept to modeling studies in oceanic biogeochemistry. It, therefore, is essential to construct the ocean carbon cycle modeling whether the Redfield stoichiometry is universal in the ocean or not. Recent model study (Klausmeler et al., 2004) suggests that the canonical Redfield N:P ratio of 16 is not a universal biochemical optimum, but instead represents an average of species-specific N:P ratios. In fact, deep-water Redfield ratios in the northern hemisphere show evidence for temporal trends over the past five decades. To elucidate the controlling factors of N:P ratios in dissolved inorganic pools is important in order to have better understanding of the ecological features of ocean as well as marine biogeochemical carbon cycling.

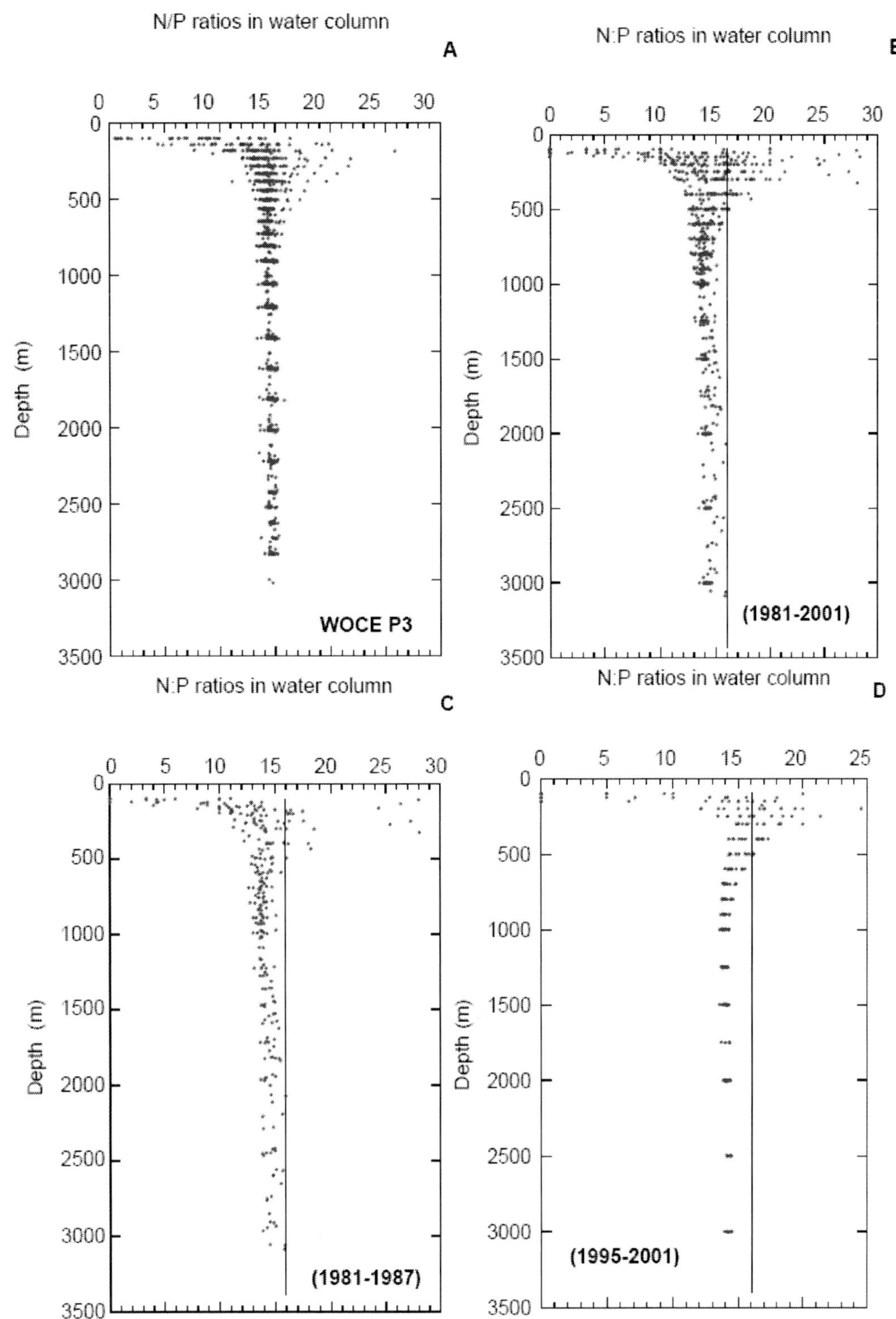

Figure 9. A composite plot of N:P ratio to depth. A: WOCE P3, B: Sta. P9-B (1981-2002), C: Sta. P9-B (1981-1987), D: Sta. P9-B (1995-2001).

The N:P ratios in dissolved inorganic nutrients were plotted as a function of depth for the WOCE P3 and Sta. P9-B time series data (1981-2001). The result is shown in Figure 9a and b. The depth plots of the N:P ratios for two cases showed the typical T shape; both larger and smaller values than the Redfield ratio appeared in shallower depth layers (100 – 500 m depth). The variation pattern of N:P ratios in WOCE P3 and time-series data differs from that for WOCE P9 (Hirose and Kamiya, 2003) and Sta. ALOHA time-series data (Karl et al., 2001), which showed the characteristic left sided T shape. Significant lower N:P ratios to the Redfield N:P ratio occurred in the shallower layers in the subtropical western North Pacific. On the other hand, the N:P ratios in deeper layers (1000 – 3000 m depth) showed relatively uniform values; there is no significant difference between the WOCE P3 and Sta. P9-B time-series data. The N:P ratios in deep layers for the WOCE P3 and Sta. P9-B time-series data are very similar to that of WOCE P9 and Sta. ALOHA time series data. However, the N:P ratios in deep waters of the subtropical North Pacific are significantly lower than the Redfield N:P ratio. The variability of deep N:P ratios is discussed in later section.

According to data analysis of GEOSECS and TTO, Fanning (1992) suggested that a depth plot of N:P ratio of the dissolved matter pool in the oligotrophic gyres of the world ocean should produce an envelope of points for a plot of N:P versus water depth that is "T" shaped as a result of increase of analytical uncertainty in low nutrient concentrations. We obtained the T-shaped profiles for the N:P ratio of inorganic nutrients of WOCE P3 and the Sta. P9-B time-series data. However, we believe that the "T" shaped profiles cannot be formed as a result of increase of analytical uncertainty in low nutrient concentrations.

A reason is that the vertical pattern of the N:P ratios in the KR region (WOCE P9) showed left sided T shape (Hirose and Kamiya, 2003), which suggests that the vertical pattern of the N:P ratios shows interannual variability such as decadal change. We plotted the N:P ratios as a function of depth for two different periods (from 1981 to 1987 and from 1995 to 2001) at Sta. P9-B. The vertical pattern of the N:P ratios in the period 1981 to 1987 showed left sided T shape except several points (Figure 9c). On the other hand, the vertical pattern of N:P ratios in the period 1995 to 2001 tended to show right sided T shape (Figure 9d).

These findings reveal that change of the N:P ratio – depth pattern cannot be generated from random errors due to larger analytical uncertainties in shallower layers.

In order to elucidate the governing factors of N:P ratios in the water column, we examined the relationship between the N:P ratios and rate constants related to marine biogeochemistry obtained from the simple model analysis of vertical profiles of nutrients in the water column. The N:P ratios in the shallower layer (175 – 200 m depth) inorganic nutrient pool are plotted to the relative consumption rate constant (k_{oP}/k_{oN}). We cannot discuss the N:P ratios in subsurface layer (100 – 150 m depth) because of larger analytical uncertainties due to very low nutrient concentrations.

The result, shown in Figure 10, suggests that the N:P ratios in the shallower layer (175 – 200 m depth) correlated to the relative consumption rate constant ($(N/P)_s = 28(k_{oP}/k_{oN}) - 13.6$). In case when nitrate is preferentially consumed by phytoplankton ($k_{oP}/k_{oN} < 1$), the N:P ratios in the shallower layer (175 – 200 m depth) are lower than the Redfield stoichiometry. We postulate that the T-shaped profiles for the N:P ratio of inorganic nutrients are produced as an overlap figure of the variation of N-like and P-like phytoplankton communities in the marine environments.

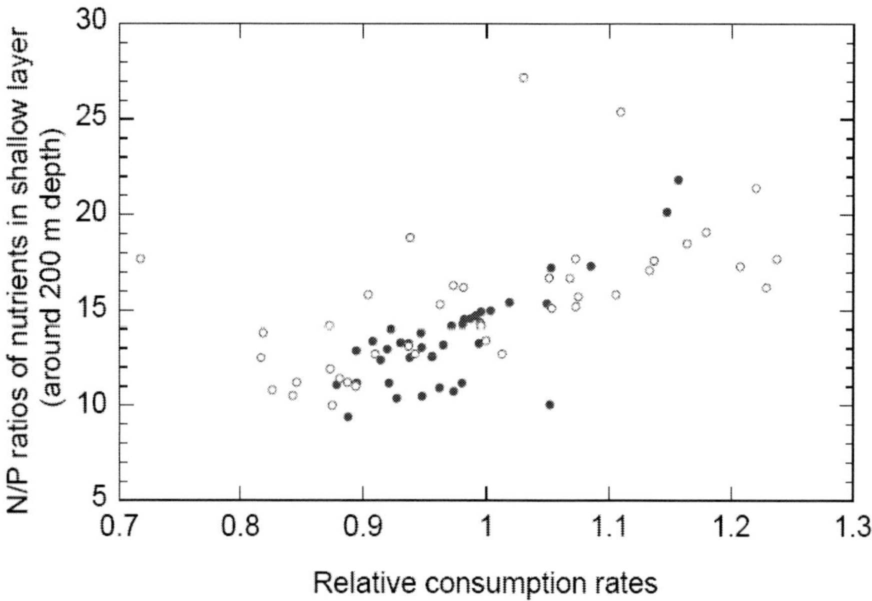

Figure 10. The relationship between N:P ratio in shallower layer (175-200 m) and the relative consumption rate (k_{oP}/k_{oN}). Open circle: Sta. P9-B, closed circle: WOCE P3.

N:P Ratios in Deep Waters

The N:P ratio of inorganic nutrients in deep water (> 1250 m) of the KR regions showed a negative offset from the Redfield ratio. A similar negative offset in deep waters observed in the oligotrophic North Pacific Ocean (GEOSECS: Fanning, 1992; Sta. ALOHA: Karl et al., 2001; WOCE P9: Hirose and Kamiya, 2003). Although the deep water N:P ratio seems to converge a single N:P ratio, there may be a significant variability of the deep water N:P ratio; ex., about ±1 % for WOCE P9 (19° – 33°N along 137°E), about ±3 % for WOCE P3 (130° – 146°E along 24° 16'N) and about ±5 % for the time-series data. The variability of the deep water N:P ratio is slightly larger than analytical uncertainties due to incomplete state of traceability. Recently, Pahlow and Riebesell (2000) suggested that deep-water Redfield ratios in the northern hemisphere show evidence for temporal trends over the past five decades based on the comprehensive data set compiled by NODC SD2. We hypothesize that the deep water N:P ratios show a natural variability. In order to elucidate factors controlling the variability of the deep water N:P ratios, the N:P ratios of inorganic nutrients in nutrient maximum layer (1250 m) were plotted by a function of the N:P ratios of the export flux at 100 m depth, which were determined from the curve-fitting coefficients determined by using the simple biogeochemical equation. The result is shown in Figure 11. The N:P ratios of inorganic nutrients in nutrient maximum layer (1250 m) are linearly related to the N:P ratios of the export flux at 100 m depth; that is, the deep water N:P ratio directly reflects the average N:P ratio of sinking particle assemblages, related to the particle export flux. The result reveals that there is no reason that the deep water N:P ratios should rigorously coincide with the Redfield stoichiometry.

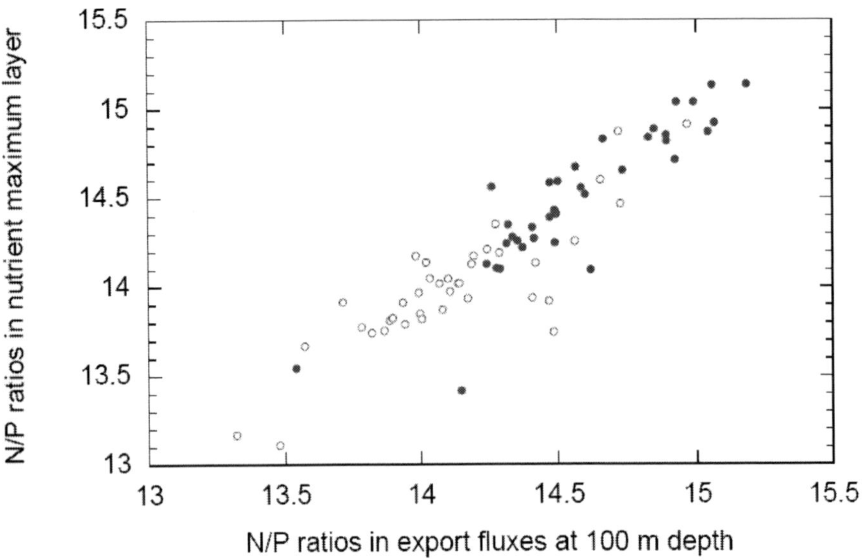

Figure 11. The relationship between N:P ratio in nutrient maximum layer (1250 m) and N:P ratio of the export flux at 100 m depth. Open circle: Sta. P9-B, closed circle: WOCE P3.

This finding suggests that the variability of surface phytoplankton assemblages leads to change of the deep water N:P ratios. Recently Rubin (2003) introduced two phytoplankton assemblages model in order to explain deep lower N:P ratios: one end member is diatoms with a lower N:P ratio of 10-13, and another is prymnesiophytes with a higher N:P ratio of >16. This curve fitting analysis with the biogeochemical function provides us to explanation of the variability of the deep N:P ratios as well as the presence of the lower N:P ratio.

Temporal Variation of N:P Ratios in Deep Waters

The N:P ratios of the export flux in the KR region reflect the N:P ratios in deep waters (>1250 m). The F_N/F_P values in Sta. P9-B are plotted as a function of time (Figure 12).

The F_N/F_P values showed relatively high value (~ 15) in the mid 1980s and decreased in the late 1980s. The F_N/F_P values in the 1990s (13.4 – 14.3) showed no long-term trend, although there is significant interannual variability of the F_N/F_P values. In order to confirm whether occurrence of relative high F_N/F_P values in the mid 1980s is natural or artifact, we superimposed the F_N/F_P values in WOCE P3 to Sta. P9-B time-series data, in which the spatial distribution of the F_N/F_P values is transformed to temporal change assuming the westward advection velocity of 770 km y^{-1} (about 2.5 cm s^{-1}) for the mesoscale eddies in the western North Pacific. The higher F_N/F_P values and their variation pattern obtained from WOCE P3 were in fair agreement with those in the Sta. P9-B time series data; i.e., the temporal variation of the F_N/F_P values reflects their spatial distribution. The result reveals that the deep sea N:P ratios in the western North Pacific show the natural temporal variability with a range of approximately ±5 % as a result of variability of phytoplankton community structure.

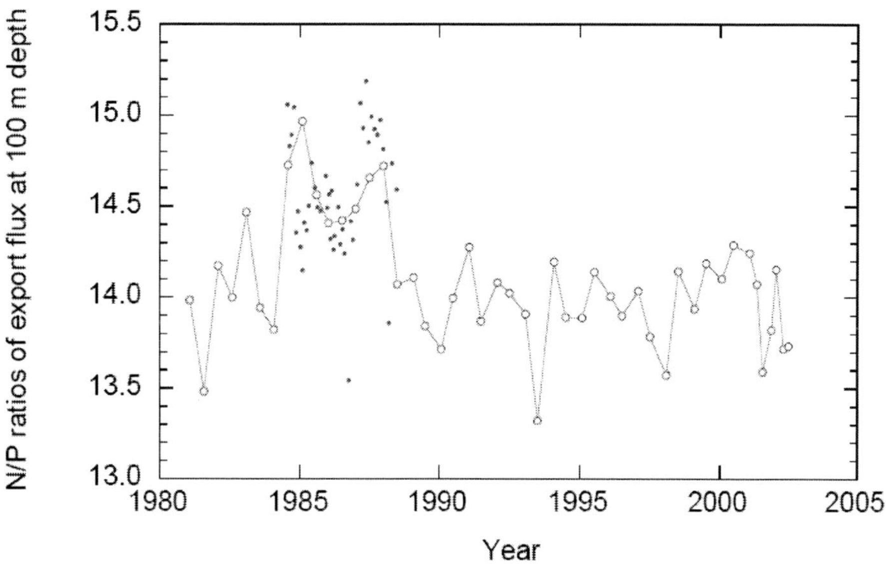

Figure 12. The temporal variation of N:P ratio of the export flux at Sta. P9-B during the period from 1981 to 2002. Open circle: Sta. P9-B, closed circle: WOCE P3.

Temporal and Spatial Variations of Relative Consumption Rates

The relative consumption rate constants are given by the ratios of B (nitrate) to B (phosphate) of the biogeochemcial curve fitting because the upwelling velocity of nitrate through the water column should essentially coincide with that of phosphate. The relative consumption rates along the WOCE P3 line ranged from 0.888 to 1.156 with an average of 0.997. In the case of $k_{oP}/k_{oN} = 1$, the consumption rate of phosphate balances to that of nitrate. Figure 13 shows the longitudinal distribution of the relative consumption rate ($=k_{oP}/k_{oN}$). The variability of the relative consumption rate for WOCE P3 was larger than that of WOCE P9 (Hirose and Kamiya, 2003). The latitudinal distribution of k_{oP}/k_{oN} values along 137°E (WOCE P9), ranging from 0.95 to 1.05, showed relatively small variation. Apparently, the consumption of nitrate in the Kuroshio and Kuroshio recirculation regions for WOCE P9 was the same as that of phosphate. The result of WOCE P9 suggested that the phytoplankton assemblage with lower relative consumption rates, which preferentially consumed nitrate, occurred in the North Equatorial Current region (Hirose and Kamiya, 2003). On the other hand, there was a larger variation of the longitudinal distribution of k_{oP}/k_{oN} values along 24° 16'N (WOCE P3) in the Kuroshio recirculation region. The N-like and P-like phytoplankton assemblages existed longitudinally as a split pattern.

The temporal variation of the relative consumption rate at Sta. P9-B is shown in Figure 14. The temporal variation of the relative consumption rate (k_{oP}/k_{oN}) ranged from 0.717 to 1.256 with an average of 1.013. The relative consumption rates were generally lower than unity during the period 1981 to 1994; phytoplankton community preferentially consumed nitrate comparing with phosphate. On the other hand, relative consumption rates after 1994 changed to be higher than unity; P-like phytoplankton assemblage was dominant in this period. This finding suggests that the relative consumption rate, which is related to marine ecosystem, exhibits long-term variability such as a regime shift.

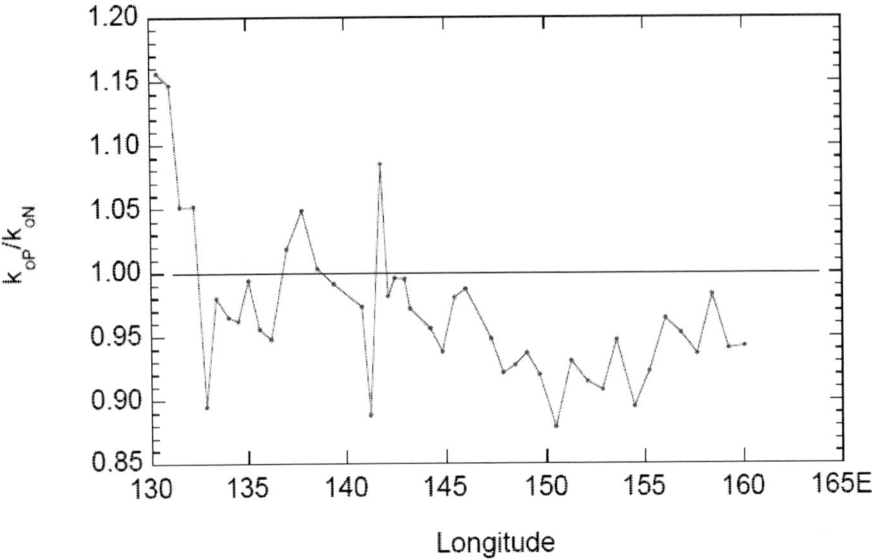

Figure 13. The longitudinal distribution of the relative consumption rate flux along 24° 16'N (WOCE P3).

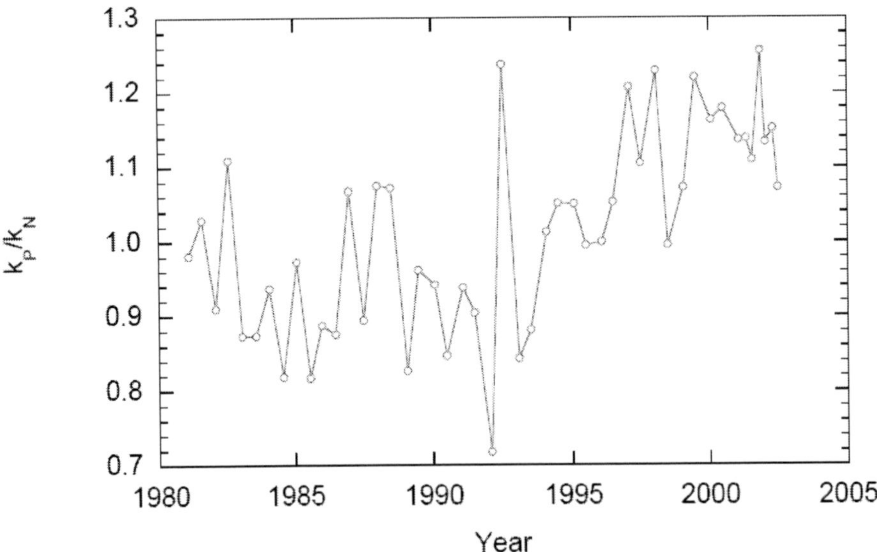

Figure 14. The temporal variation of the relative consumption rate at Sta. P9-B during the period from 1981 to 2002.

In fact, the regime shift of marine ecosystem in the eastern North Pacific occurred in 1989 accompanied with the climate changes from 1989 to 1977 (Mantua, 2004). The result may suggest that a kind of regime shift in the western North Pacific ecosystem occurred around 1993 although the change of the relative consumption rate did not coincide with climatological regime shift in 1988/89 (Yasunaka and Hanawa, 2002).

There is a question what physical or climatological processes lead to temporal variation of the relative consumption rate of nutrients in the KR region. The decadal variability of sea surface temperature (SST), which results in atmosphere-ocean interaction, occurred in the North Pacific. The SST near Japan showed an increasing trend in the late 1990s. The SST trend is similar to that of the relative consumption rate. The increase of SST may lead to ecological change of phytoplankton community from N-like to P-like species.

The Ecological Role of Cyclonic Eddies

The winter convection due to cooling and strong wind is one of dominant processes to transport deeper nutrient-rich water to surface layer. However, this process cannot continuously support oceanic biomass because most of the nutrients in surface layer are consumed by spring blooming by phytoplankton. Eddy pumping (Jenkins, 1988; Falkowski et al., 1991; Garçon et al., 2001) can continuously supply nutrient-rich water to euphotic zone during life-time of eddies. Therefore, meso-scale physical dynamics such as production of eddy and its motion plays a significant role for ocean ecosystem and biogeochemical carbon cycling. The cyclonic eddies in the western North Pacific, which are generated in the vicinity of the Kuroshio extension region in winter (Ebuchi and Hanawa, 2001), are able to continuously supply nutrients from deep waters to the euphotic layer. In case when the meso-scale eddy has a diameter of about 200 km and an average upwelling velocity of 5 m y^{-1}, we have 0.71 $x10^{10}$ mol N y^{-1} as the nitrogen export flux in a cyclonic eddy, which corresponds to the carbon export flux of about 4.7 $x10^{10}$ mol C y^{-1}. (0.56 Mtons C y^{-1}). Although the calculated carbon export flux in the cyclonic eddy is lower than highly productivity area such as the Subarctic Pacific, it is greater than that in the surrounded oligotrophic ocean. Another important aspect is that the cyclonic eddies with a significant amount of biomass including nekton moves southwestward in the KR region of the western North Pacific. Most of the eddies disappeared at the northwestern edge of the Philippine basin (Ebuchi and Hanawa, 2001) in which corresponds to upstream of the Kuroshio Current. The Kuroshio Extension region, generating mesoscale eddies due to meander of the Kuroshio Current, corresponds to highly productivity area in the North Pacific, whereas the Kuroshio Current contains areas spawned by fishes such as sardines. This means that the westward movement of cyclonic eddies can transport a significant amount of biomass from the Kuroshio Extensions region to upstream of the Kuroshio Current. This process may enhance reproduction of biomass such as nekton in the western North Pacific. The variability of mesoscale physical dynamics due to climate change, variability of generation of eddies and their transport, significantly affects marine ecosystem.

Quality of Nutrient Data

As a result of analysis of nutrient data in the ocean, the natural variability of the nutrient concentrations is less than 5 % of average values in deep waters. Therefore, high quality dataset of nutrient concentrations with high precision (accuracy: 1 %, precision: 0.2 %) has been required (WOCE, 1994). Especially, in order to detect the temporal and spatial variability of nutrients in the ocean, it is necessary to ensure comparability of nutrient data between cruses. Unfortunately, the comparability of nutrient data in the ocean has not been established in the field of oceanography because we have no appropriate standard materials of nutrients in seawater. The curve fitting technique for the vertical profile of nutrients in the water column can eliminate random error of individual analytical value, however, cannot remove unexpected offset of nutrient data between cruises. Therefore, we cannot exclude the possibility that the offset of nutrient analysis affects variability of vertical profiles of nutrient in the ocean. To remove offset of nutrient data, it is needed to introduce certified reference materials of nutrient analysis. Recently, a lot of effort has been paid to develop reference materials of nutrients in seawater and to conduct intercomparison exercise of nutrient analysis. Intercomparioson studies clearly show that global use of reference materials of nutrients in seawater would greatly improve the comparability of nutrients data in the world's oceans (Aoyama et al., 2007). As a result, a high quality data set of nutrient concentrations in the ocean (comparability: <0.5 %; precision: <0.2 %) has been reported (Aoyama et al., 2008). In the near future, more detailed changes of the nutrient concentrations in the ocean will be detectable.

CONCLUSION

The temporal and spatial variability of nutrients profiles in the Kuroshio recirculation (RE) region of the western North Pacific was examined based on the simple biogeochemical curve fitting method, which can be well reproduced the vertical profiles of nitrate and phosphate from 100 m to 2500 m depth (Hirose and Kamiya, 2003). The major physical process governing the variability on the vertical profiles of the nutrient concentrations in the KR region is eddy-driven upwelling and downwelling. The generation and transport of cyclonic and anticyclonic eddies affected by climate change lead to the variability of nutrient supply to euphotic zone. This phenomenon may give a marked impact to the marine ecosystem.

The vertical profile of nutrients in the water column is controlled by some biogeochemical processes as do the physical processes. The N:P ratio of inorganic nutrient pool is a key factor to solve how biogeochemical processes lead to the variability of nutrients in water columns. The curve fitting analysis with the biogeochemical function revealed that the N:P ratio in shallower layer is directly related to the relative consumption rate of nutrients (= ratio of P consumption rate to N consumption rates); the relative consumption rate of >1 means preferential consumption of nitrate comparing with phosphate. The temporal and spatial variability of N:P ratio in shallower layer is attributable to the change of the relative consumption rate which is due to the change of dominant phytoplankton species.

In deep waters, there is natural variability of the N:P ratios with an amplitude of less than 5 %. The N:P ratio in deep waters (> 1000 m) is linearly related to that in export particles (= N export flux/P export flux from surface layer). Therefore, the N:P ratio in deep waters is affected by the variability of the composition of phytoplankton communities in surface waters; especially, lower N:P ratio in deep waters occurs when dominant species of export particles is diatoms.

Climate change affects significantly mesoscale physical dynamics (ex., eddy production) of the North Pacific Ocean. The change of eddy production in the Kuroshio Extension region results in variation of nutrient supply to the euphotic zone and nutrient composition (N:P ratio). The change of nutrient dynamics may lead to the variability of marine ecosystem and carbon cycling in the North Pacific.

ACKNOWLEDGMENT

The author thanks the staff members of Geochemical Research Department for constructive comments and suggestion.

REFERENCES

Anderson, L.A.. and Sarmiento, J.L. (1994). Redfield ratios of remineralization determined by nutrient data analysis. *Global Biogeochem. Cycles*, 8, 65-80.

Aoyama, M., Takeuchi, A., Seike, T., and Miyabe, S. (2008) *Nutrients,* MR0704 Cruise Report, JAMSTEC.

Aoyama, A. Becker, S., Dai, M., Daimon, H., Gordon, L.I., Kasai, H., Kerouel, R., Kress, K., Masten, D., Murata, A., Nagai, N., Ogawa, H., Ota, H., Saito, H., Saito, K., Shimizu, T., Takano, H., Tsuda, A., Yokouchi, K., and Youenou, A. (2007). Recent comparability of oceanographic nutrients data: result of a 2003 intercomparison exercise using reference materials. *Analytical Sciences*, 23, 1151-1154.

Broecker, W.S., and Peng, T.-H. (1972). *Tracers in the Sea*. Eldigio Press, Palisades, New York, 690 pp.

Broecker, W.S., Takahashi, T., and Takahashi, T. (1985). Sources and flow patterns of deep-ocean waters as deduced from potential temperature, salinity, and initial phosphate concentration. *Journal of Geophysical Research*, 90, 6925-6933.

Dore, J.E., and Karl, D.M. (1996). Nitrite distributions and dynamics at Station ALOHA, *Deep-Sea Research, Part II*, 43, 385-402.

Ebuchi, N., and Hanawa, K. (2000). Mesoscale eddies observed by TOLEX-ADCP and TOPEX/POSEIDON altimeter in the Kuroshio Recirculation region south of Japan. *Journal of Oceanography*., 56, 43-57.

Ebuchi, N., and Hanawa, K. (2001). Trajectory of mesoscale eddies in the Kuroshio Recirculation region. *Journal of Oceanography.,* 57, 471-480.

Falkowski, P.G.. (1997). Evolution of the nitrogen cycle and its influence on the biological sequestration of CO_2 in the ocean. *Nature,* 387, 272-275.

Falkowski, P.G., Ziemann, D., Kolber, Z., and Bienfang, P.K. (1991). Role of eddy pumping in enhancing primary production in the ocean. *Nature*, 352, 55-58.

Fanning, K.A. (1992). Nutrient provinces in the sea: concentration ratios, reaction rate ratios and ideal covariation. *Journal of Geophysical Research*, 97, 5693-5712.

Fenchel, T., King, G.M., and Blackburn, T.H. (1998). Bacterial Biochemistry. In *The Ecophysiology of Mineral Cycling*, Second Edition, Academic Press.

Flierl, G.R. (1981). Particle motions in large-amplitude wave field, Geophys. *Astrophys. Fluid Dynamics*, 18, 39-74.

Franks, P.J.S., Wroblewski J.S., and Flierl, G.R. (1986). Prediction of phytoplankton growth in response to the frictional decay of a warm-core ring. *Journal of Geophysical Research*, 91, 7603-7610.

Ganachaud, A. and Wunsch, C. (2000). Improved estimates of global ocean circulation, heat transport and mixing from hydrographic data. *Nature*, 408, 453-456.

Garcçn, V.C., Oschlies, A., Doney, S.C., McGillicuddy , D., and Waniek, J. (2001). The role of mesoscale variability on plankton dynamics in the North Atlantic. *Deep-Sea Research, part II.*48, 2199-2226.

Goldman, J.C., McCarthy, J.J., and Peavey, D.G., (1979). Growth rate influence on the chemical composition of phytoplankton in oceanic waters, *Nature*, 279, 210-215.

Gruber, N., and Sarmiento, J.L. (1997). Global patterns of marine nitrogen fixation and denitrification, *Global Biogeochemical Cycles*, 11, 235-266.

Hirose, K., and Kamiya, H. (2003). Vertical nutrient distributions in the western North Pacific: Simple model for estimating nutrient upwelling, export flux and consumption rates. *Journal of Oceanography, 59,149-161.*

Japan Meteorological Agency (1995) *Cruise report of WOCE P9 one-time survey.*

Jenkins, W.J. (1988). Nitrate flux into the euphotic zone near Bermuda. *Nature,* 331, 521-523.

Kaneko, I., Takatsuki, Y., Kamiya, H., and Kawae, S. (1998). Water-property and current distributions along the WHP-P9 section (137-142°E) in the western North Pacific. *Journal of Geophysical Research*, 103, 12,959-12,986.

Karl, D.M., Christian, J.R., Dore, J.E., Latelier, R.M., Tupas, L.M., C.D. and Winn, C.D. (1996). Seasonal and interannual variability in primary production and particle flux at station ALOHA. *Deep-Sea Research*, 43, 539-568.

Karl, D.M., Letelier, R.M., Tupas, L.M., Dore, J.E., Christian, J.R., and Helbel, D.V. (1997). The role of nitrogen fixation in biogeochemcal cycling in the subtropical North Pacific Ocean. *Nature,* 388, 533-538.

Karl, D.M., Bjorkman, K.M., Dore, J.E., Fujieki, L., Hebel, D.V., Houlihan, T., Latelier, R.M., and Tupas, L.M., (2001). Ecological nitrogen-to-phosphorus stoichiometry at station ALOHA. *Deep-Sea Research, part II*, 48, 1529-1566.

Kishi, M.J. (1994). Prediction of phytoplankton growth in a warm-core ring using three dimensional ecosystem model. *Journal of Oceanography*, 50, 489-498.

Klausmeier, C.A., Litchman, E., Daufresne, T., and Levin, S.A. (2004). Optimal nitrogen-to-phosphorus stoichiometry of phytoplankton, *Nature*, 429, 171-174.

Laws, E.A., and Bannister, T.T. (1980). Nutrient- and light-limited growth of Thalassiosira fluviatilis in continuous culture, with implications for phytoplankton growth in the ocean, *Limnology and Oceanography*, 25, 457-473.

Maier-Reimer, E. (1993). Geochemical cycles in an ocean general circulation model. Preindustrial tracer distributions. *Global Biogeochemical Cycles*, **7**, 645-677.

Mantua, N. (2004). Methods for detecting regime shifts in large marine ecosystems: a review with approaches applied to North Pacific data, *Progress in Oceanography*, 60, 165-182.

McGillicuddy Jr., D.J., Robinson, A.R., Siegel, D.A., Jannasch, H.W., Johnson, R., Dickey, T.D., McNeil, J., Michaels, A.F., and Knap, A.H. (1998). Influence of mesoscale eddies on new production in the Sargasso Sea. *Nature*, 394, 263-266.

Pahlow, M., and Riebesell, U. (2000). Temporal trends in deep ocean Redfield ratios, *Science*, 287, 831-833.

Oschlies, A., and Garcçn, V.C. (1998). Eddy-induced enhancement of primary production in a model of the North Atlantic Ocean. *Nature,* 394, 266-269.

Redfield, A.C., Ketchum, B.H., and Richards, F.A. (1963). The influence of organisms on the composition of seawater. In: Hill, M.N. (Ed.), *The Sea, Ideas and Observations on Progress in the Study of the Seas*, Vol. 2 Inerscience, New York, pp. 26-77.

Rubin, S.I. (2003). Carbon and nutrient cycling in the upper water column across the Polar Frontal Zone and Antarctic Circumpolar Current along 170W. *Global Biogeochemical Cycles,* 17, 1087, doi:10.1029/2002GB001900, 2003.

Sakamoto, C.M., Karl, D.M., Jannasch, H.W., Bidigare, R.R., Letelier, R.M., Walz, P.M., Ryan, J.P., Polito, P.S., and Johnson, K.S. (2004). Influence of Rossby waves on nutrient dynamics and the plankton community structure in the North Pacific subtropical gyre. *Journal of Geophysical Research*, 109, C05032, doi:10.1029/2003JC001976.

Tyrrell, T. (1999). The relative influences of nitrogen and phosphorus on oceanic primary production, *Nature*, 400, 525-531.

Van Cappellen, P., and Ingall, E.D. (1994). Benthic phosphorus regeneration, net primary production, and ocean anoxia: A model of the coupled marine biogeochemical cycles of carbon and phosphorus, *Paleoceanography*, 9, 677-692.

Watson, K.J., and Whitfield, M. (1985). Composition of particles in the global. *Deep-Sea Research,* 32, 1023-1039.

WOCE (1994). WOCE operational manual, WHP Office Report WHPO 91-1, *WOCE Report* No. 68/91, Woods Hole, USA.

WOCE (2007). WOCE Atlas Volume 2. Pacific Ocean http://www. woce.org/atlas_webpage/

Yamanaka, Y., and Tajika, E. (1996). The role of the vertical fluxes of particulate organic matter and calcite in the oceanic carbon cycle: Studies using an ocean biogeochemical general circulation model. *Global Biogeochemical Cycles*, 10, 361-382.

Yasunaka, S., and Hanawa, K. (2000). Regime shifts found in the northern hemisphere SST field. *Journal of the Meteorological Society of Japan*, 80, 119-135.

In: The Pacific and Arctic Oceans
Editor: Kallen B. Tewles

ISBN: 978-1-60692-010-7

Chapter 3

COMPARISON OF MESOZOOPLANKTON BIOMASS DOWN TO THE GREATER DEPTHS (0-3000 M) BETWEEN 165°E AND 165°W IN THE NORTH PACIFIC OCEAN: THE CONTRIBUTION OF LARGE COPEPOD *NEOCALANUS CRISTATUS*

Atsushi Yamaguchi*
Laboratory of Marine Biology, Graduate School of Fisheries Sciences,
Hokkaido University, 3-1-1 Minatomachi, Hakodate,
Hokkaido 041-8611, Japan

ABSTRACT

While our knowledge about east-west differences in seasonal features of phytoplankton communities in the North Pacific ecosystem has advanced quickly in recent years (i.e. presence of phytoplankton spring bloom in the west while it absent in the east), little information is available on east-west differences in organisms at higher trophic level than phytoplankton. This is partly due to the differences in sampling methodologies for animals at higher trophic levels. In the present study, we evaluate east-west differences of zooplankton biomass down to the greater depths in the North Pacific employing the same method. We carried out zooplankton sampling with VMPS (60 μm mesh) down to the greater depths (six layers between 0 and 3000 m) at 19 stations between 36°N and 50°N along 165°E and 165°W in summers of 2003 and 2004. Half of the samples were filtered on 30 μm mesh and used for biomass determination. The other half samples were preserved with 5% borax-buffered formalin and used for microscopic observation. In the land laboratory, biomass determination was done for 8 mass-units (WM, DM, C, H, N, ash, AFDM and Energy). As taxonomic account, large calanoid copepod *Neocalanus cristatus* CV stage were enumerated and their biomass was determined by multiplying separately determined individual mass. Zooplankton biomass integrated over the 0-3000 m depth varied from 5.9 to 28.0 g DM m^{-2}, and was higher at high latitudes. From the viewpoint of east-west comparison, zooplankton biomass in the

* Fax: +81-138-40-5542; E-mail: a-yama@fish.hokudai.ac.jp

western (165°E) stations was 1.7 times higher than that in the eastern (165°W) stations consistently. As a taxonomic account, biomass of *N. cristatus* CV was greater at western stations (165°E) than at eastern stations (165°W), which reflected to the east-west differences in the whole zooplankton biomass. Carbon contents of *N. cristatus* CV individuals were higher for the individuals from 165°E stations than those from 165°W stations. These east-west differences in zooplankton biomass (high in west, low in east) may be interpreted by regional differences in phytoplankton abundance of both regions (high in west, low in east).

INTRODUCTION

In the pelagic ecosystem, mesozooplankton feed on phytoplankton, microzooplankton, other mesozooplankton and detritus. Since mesozooplankton produces large-sized fecal pellets, size and biomass of them are known to be primary important to determine amount of vertical particle flux (cf. Michaels and Silver 1988). Together with their contribution of vertical material flux, their contribution on remineralization of organic matter is also important. Hernández-León and Ikeda (2005) estimate global depth-integrated mesozooplankton respiration is 13.0 Gt C year^{-1}, which accounts to 17-32% of global primary production. From viewpoint of fisheries, mesozooplankton is a major food source of epipelagic fishes. Size and biomass of mesozooplankton effects on food selectivity of epipelagic fishes (Sheldon et al. 1977) and taxonomic (=chemical) contents of mesozooplankton effects on growth and survival of larvae (van der Meeren and Næss 1993). Thus, accurate estimation of biomass and chemical composition of mesozooplankton is a prime important from the viewpoints both biological oceanography and fisheries.

In the last decade, east-west differences of plankton community in the North Pacific are revealed. Thus, phytoplankton density and primary production are known to be higher in the western than those in the eastern North Pacific (Shiomoto and Asami 1999, Shiomoto and Hashimoto 2000). This east-west difference is considered to be caused by the differences in concentration of dissolved-iron (higher in the western) (Harrison et al. 1999, Suzuki et al. 2002). East-west differences in phytoplankton community assumed to effects zooplankton abundance, biomass, and community in these region (Mackas and Tsuda 1999), and higher transfer efficiency in the western region is suggested (Taniguchi 1999). For the dominant copepods in mesozooplankton community, east-west differences in life cycle (Kobari and Ikeda 2001, Padmavati et al. 2004, Tsuda et al. 2004, Shoden et al. 2005) and their body sizes (Tsuda et al. 2001, Kobari et al. 2003) are reported. However, there still remains whether the total mesozooplankton masses and chemical contents are varied between east and west or not. It is partly because of the differences in sampling device (i.e. mesh size, sampling period, sampling depth etc.) between east and west, which prevents direct comparison of biomass and chemical composition of mesozooplankton between east and west.

In the present study, mesozooplankton biomass and their chemical composition are quantified based on same method from sea surface to 3000 m at 36°N-50°N along 165°E (western) and 165°W (eastern) North Pacific. Addition to biomass and chemical components, as a taxonomic account, biomass and body size of large sized copepod *Neocalanus cristatus* CV in the samples are quantified to examine regional changes in taxonomic accounts and evaluate their effects on total mesozooplankton biomass.

Material and Methods

Field Sampling

Zooplankton samples were collected at 19 stations between 36°N and 50°N along 165°E and 165°W transects in the North Pacific Ocean during 1 July to 2 August of 2003 and 2004 (Figure 1). There were four (2003) or five (2004) stations along 165°E and three (2003) or seven (2004) stations along 165°W (Figure 1). At each station, stratified vertical sampling from six strata (0-50, 50-250, 250-500, 500-1000, 1000-2000 and 2000-3000 m) were made with 60-µm mesh and 0.25 m^2 (= 0.5 m x 0.5 m) mouth area Vertical Multiple Plankton Sampler (VMPS, Tsurumi Seiki Ltd., Terazaki and Tomatsu 1997) at night (stratified sampling by VMPS took ca. 3 hours and was made during 19:28-3:48 in local time). Towing speed of VMPS was 1 m s^{-1}, and the TSK flowmeter was mounted of the mouth of VMPS to monitor volume of filtered water. The filtering efficiency was mostly ca. 90%. At each station, CTD cast (SBE 19 plus) was also made down to 3000 m.

Collected zooplankton samples were split with three fractions (1/2, 1/4 and 1/4) by Motoda Splitting device (Motoda 1959). A 1/2 of the samples was preserved with 5% borax buffered formalin-seawater for latter microscopic observations, and the 1/4 aliquot was filtered through pre-weighed pieces of 30 µm mesh net under vacuum. Samples were then briefly rinsed with distilled water while still under vacuum, wrapped with aluminum foil, placed into sealed plastic bags, and stored in a deep freezer (-80°C).

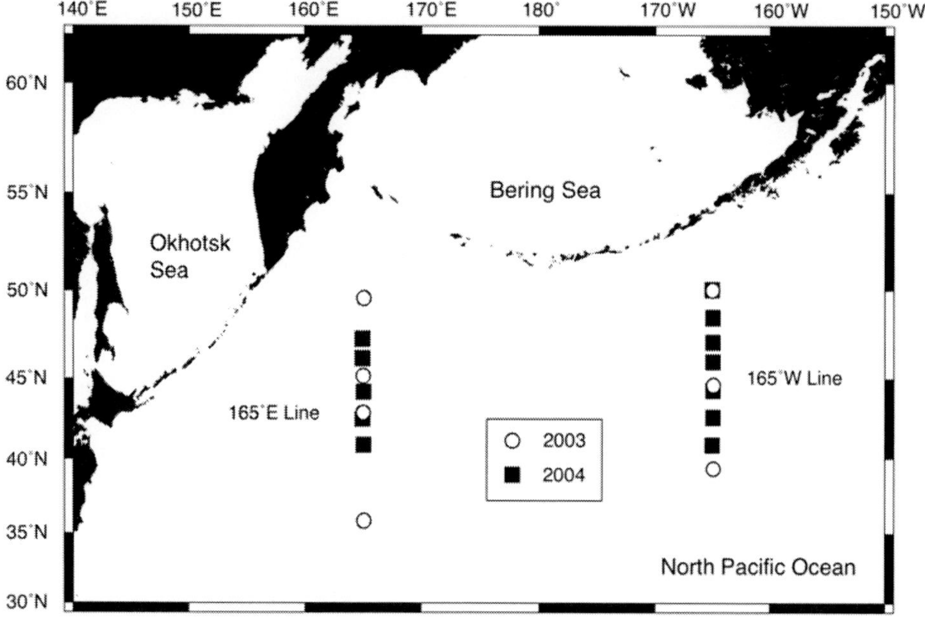

Figure 1. Location of sampling stations along 165°E and 165°W lines in the North Pacific Ocean in 2003 (open circle) and 2004 (solid square). There were four (165°E) and three (165°W) stations in 2003, while were five (165°E) and seven (165°W) stations in 2004.

The remaining 1/4 aliquot was used to sort fresh zooplankton especially copepod *Neocalanus cristatus* copepdid stage V (CV). Sorted *N. cristatus* CV was briefly rinsed with distilled water, placed on pre-weighed aluminum pan and also stored in a deep-freezer.

Chemical and Taxonomic Component

In the land laboratory, frozen zooplankton and *Neocalanus cristatus* CV samples were weighed (wet mass, WM) and then freeze-dried to obtain dry mass (DM). Water content in the samples was calculated from the difference between WM and DM, and expressed as a percentage of WM. Dried samples were ground into a fine powder with a ceramic mortar and pestle and used for analyses of carbon and nitrogen (Yanaco, CHN Corder MT-5). A weighed fraction of each sample was put into pre-weighed aluminum cup, then incinerated at 480°C for 5 h and re-weighed for ash determination (ash), and ash-free dry mass (AFDM) fraction was calculated (AFDM= DM- ash). The energy content was calculated by using formulae given by Gnaiger (1983), amended by Gnaiger and Shick (1985): $J= 66.265W_C+ 4.436W_N- 11.2$, where J is an energy content in J mg^{-1} AFDM, and W_C and W_N are fractions of C and N, respectively, on an AFDM basis.

As a taxonomic component, *Neocalanus cristatus* CV were counted with the formalin preserved samples. Using individual DM data at each station, biomass of *N. cristatus* CV were quantified (g DM m^{-2}) and expressed as percentage to total DM. The prosome length (PL) of *N. cristatus* CV in 2003 was measured to the nearest 0.2 mm under a dissecting microscope with an eye-piece micrometer.

Analysis of Biomass and Chemical Components

For chemical component analysis, a cluster analysis and non-metric multidimensional scaling (NMDS) ordination were made. Data (X) on masses (WM, DM, C, N, Ash and AFDM, all in mg m^{-3}), energy (J m^{-3}), and chemical contents (Water [%WM], C [%DM], N [%DM], C:N ratio [weight base], AFDM [%DM] and Energy [J mg^{-1} DM]) at each stratum were transformed to $\log_{10}(X+1)$ prior to analysis in order to reduce the bias of chemical components.

Similarities between samples were examined by the Bray-Curtis index (Bray and Curtis 1957) according to the differences in masses, energy and chemical components. For grouping the samples, the similarity indices were coupled with hierarchical agglomerative clustering with a complete linkage method. The NMDS ordination was carried out to delineate the sample groups on the two-dimensional map. All of these analyses were carried out using BIOSTAT II software (Sigma Soft). To evaluate differences in biomass and chemical composition between groups, one-way ANOVA and post-hoc test with Fisher's PLSD were employed.

Results

Hydrography

Sea surface temperature varied from 7.9°C (49°30'N, 165°00E in 2003) to 20.3°C (36°00'N, 165°00'E in 2003) (Figure 2). Both 2003 and 2004, the 6°C isothermal contours down to 250 m around at 42°30'N in the 165°E, while they down to 250 m around at 46°00'N in the 165°W. Thus, temperature conditions in 165°W were warmer than those in 165°E. Both 165°E and 165°W, temperature decreased with increasing depth, and was <3°C below 1000 m throughout the region.

Salinities varied from 32.4 PSU (50°N, 165°W in 2004) to 34.5 PSU (36°00'N, 165°00'E in 2003) (Figure 2). Both 2003 and 2004, the 33.6 PSU isoline reached sea surface around at 42°30'N in the 165°E, while they around at 41°00'N in the 165°W. Thus, salinity conditions near surface in 165°W were less saline than those in 165°E. Both 165°E and 165°W, salinity increased with increasing depth, and was >34.5 below 1500 m throughout the region.

Based on criteria of Favorite et al. (1976), subarctic front (SF, northern end of transition domain) and subarctic boundary (SB, southern end of transition domain) were identified. Both 2003 and 2004, the SF located at around 44°N in the 165°E, while was at around 47°N in the 165°W (Figure 2).

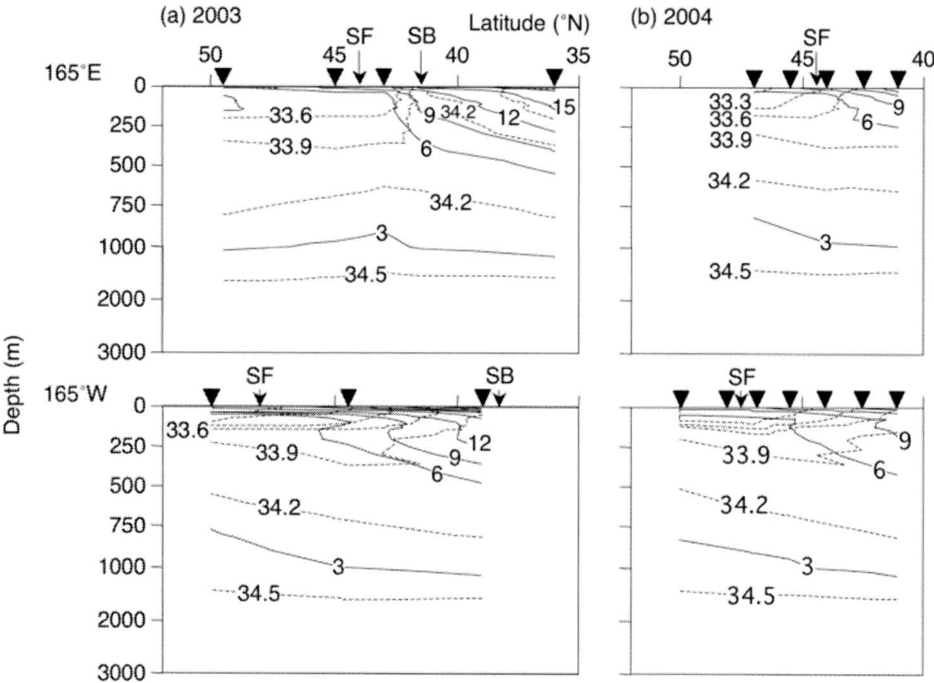

Figure 2. Vertical structures of temperature (solid line) and salinity (dashed line) along 165°E (upper) and 165°W (lower) in the North Pacific Ocean in 2003 (a) and 2004 (b). Note that the depth intervals are varied between upper and lower of 1000 m depth. Solid triangles indicate sampling stations. Approximate positions of subarctic front (SF) and subarctic boundary (SB) are marked with arrows.

In 2003, SB detected at 42°N in 165°E, while was at 38°N in 165°W. Note that SB was not detected in 2004, because of the limited latitudinal study area (41°N-50°N) in this year. Thus the northern end of transition domain (SF) in 165°W was located northward than that in 165°W.

Zooplankton Mass and Chemical Composition

Zooplankton dry masses throughout the 0-3000 m depths were varied between 5.9 g DM m^{-2} (39°00'N, 165°W in 2003) and 28.0 g DM m^{-2} (47°00'N, 165°E in 2004) (Figure 3). Both 2003 and 2004, zooplankton mass was greatest in the highest latitude stations and was decreased with decreasing latitudes. Compare within the same latitude, zooplankton mass was consistently greater in 165°E than that in 165°W, and this phenomenon was also the case of other mass unit (data not shown). Factor between masses in 165°E and 165°W (values in 165°E: 165°W) was varied between 1.4-1.9 (mean±1sd: 1.7±0.2).

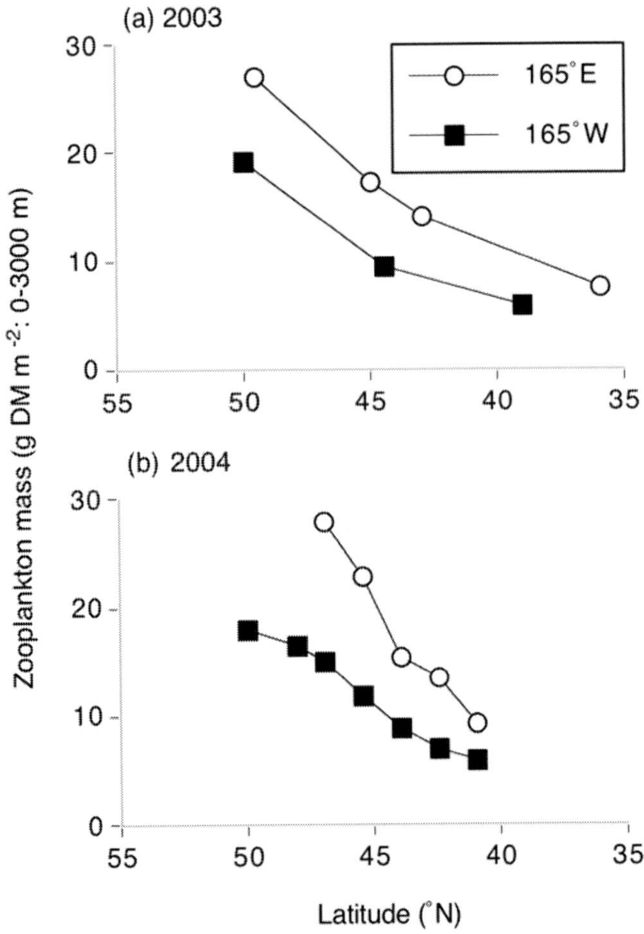

Figure 3. Latitudinal changes in zooplankton mass throughout 0-3000 m water column along 165°E (open circle) and 165°W (solid square) in the North Pacific in 2003 (a) and 2004 (b).

Cluster analysis based on the masses and chemical contents of each sample showed that zooplankton communities could be divided into five groups (A-E) at 18% Bray-Curtis dissimilarity (Figure 4). Two-dimensional NMDS plots showed that these groups were well separated from each other (Figure 4). As an environmental parameter, significant multiple regression with the NMDS ordination scores was observed for depth (r^2=0.61, p<0.001) and latitude (r^2=0.07, p<0.05). The directions of the arrows indicate that groups A-D occurred in the order of from shallow to deep: e.g. A<B<C<D (Figure 4).

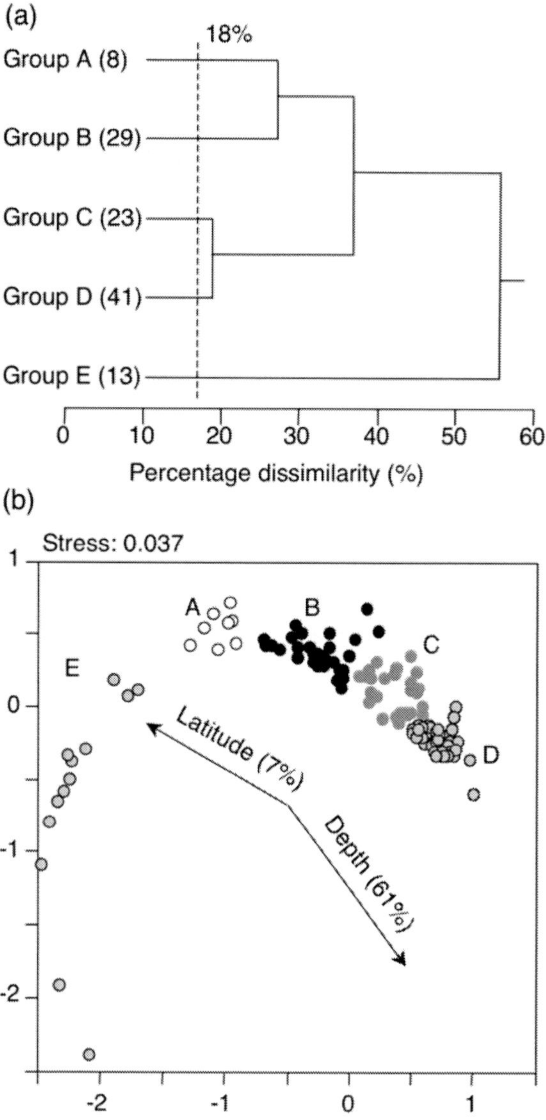

Figure 4. (a): Cluster diagram of samples based on zooplankton masses and chemical contents. Five groups (A-E) were identified at 18% dissimilarity (dashed line). Numbers in the parentheses indicate sample number each group contains. (b): NMDS ordination of samples with the five zooplankton groups derived by cluster analysis. Arrows indicate multiple regressions between ordination scores and environmental variables (p<0.05), squared multiple r were shown as percentages.

Latitudinal and vertical distribution of each zooplankton community showed that the group D occurred in the bathypelagic layer (1000-3000 m) in most of the stations (Figure 5).

Group B and C mainly occurred in the mesopelagic layer (200-1000 m) and commonly the group B occurred shallower than the group C. In the epipelagic layer (0-200 m), the shallowest depth (0-50 m) was dominated by group A and E. Distribution of group A and E separated latitudinal: i.e. the group E was seen mainly in the northern area (ca. >43°N) while the group A in the southern area (<43°N) (Figure 5).

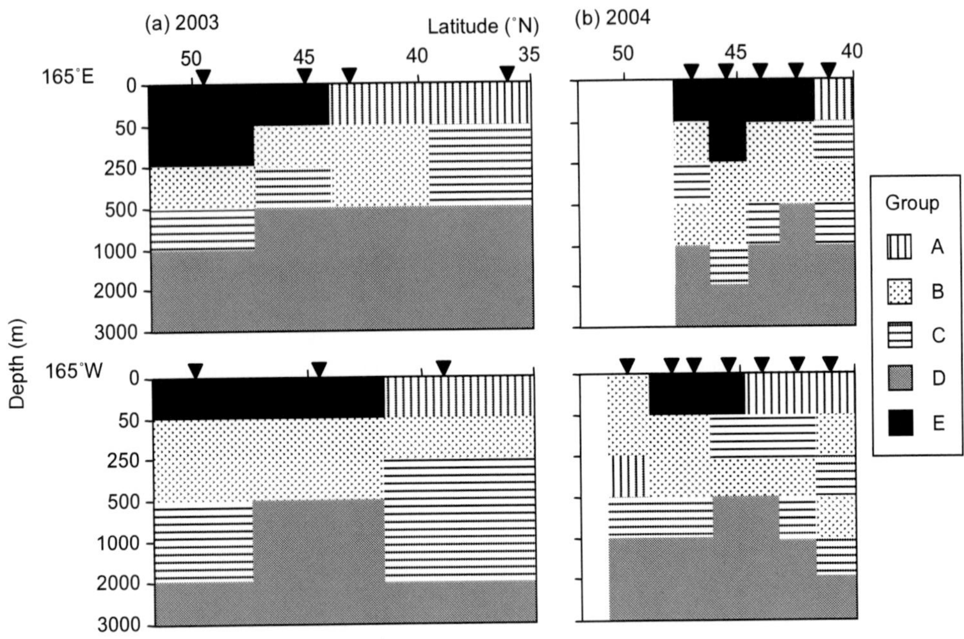

Figure 5. Latitudinal and vertical distribution of five zooplankton community groups (A-E) identified by Bray-Curtis similarity index data (cf. Figure 4) at 165°E (upper) and 165°W (lower) in 2003 (a) and 2004 (b). Triangles indicate sampling stations.

Most of the masses, energy and chemical contents showed significant differences between groups (Table 1). Masses and energy showed similar pattern, thus the group E was characterized by significantly greater than the other groups.

Values in masses and energy was in the order of E>A>B>C>D, which was same to the order of vertical distribution of each group from shallow to deep (cf. Figure 5).

Chemical contents of group E and A were comparable. Thus, the group E characterized by low Water, high C, N, AFDM and Energy contents, while the group A characterized by high Water, low C, N, AFDM and Energy contents (Table 1).

Taxonomic Component (*Neocalanus Cristatus*)

Biomass of *Neocalanus cristatus* CV varied between 0.07 g DM m^{-2} (36°00'N, 165°00'E in 2003) and 6.4 g DM m^{-2} (47°00'N, 165°00'E in 2004) (Figure 6).

Table 1. Masses, energy and chemical contents of zooplankton samples in the North Pacific down to 3000 m depths

	Group					one-way	Post-hoc test by Fisher's PLSD				
	A (n=6)	B (n=30)	C (n=23)	D (n=41)	E (n=14)	ANOVA					
Masses (mg m^{-3})											
Wet (WM)	289±64	114±44	45±18	11±6	1026±967	***	D	C	B	A	E
Dry (DM)	31.4±6.2	13.1±4.4	5.0±1.7	1.3±0.7	126.3±99.6	***	D	C	B	A	E
Carbon (C)	11.09±1.77	5.67±1.70	2.12±0.75	0.57±0.34	56.36±42.07	***	D	C	B	A	E
Nitrogen (N)	2.25±0.48	1.07±0.37	0.41±0.16	0.11±0.07	10.07±6.92	***	D	C	B	A	E
Ash (Ash)	9.96±5.33	2.82±1.65	1.12±0.56	0.26±0.13	22.85±29.42	***	D	C	B	A	E
Ash-free dry (AFDM)	21.5±3.5	10.3±3.4	3.9±1.4	1.0±0.6	103.4±75.7	***	D	C	B	A	E
Energy (J m^{-3})	504±81	265±77	99±35	26±16	2620±1973	***	D	C	B	A	E

Table 1. (Continued)

	Group					one-way	Post-hoc test by Fisher's PLSD				
	A (*n*=6)	B (*n*=30)	C (*n*=23)	D (*n*=41)	E (*n*=14)	ANOVA					
Chemical content											
Water (%WM)	88.5±4.6	87.8±3.0	88.2±3.1	89.8±2.6	86.9±3.1	**	E	B	C	A	D
C (%DM)	36.0±6.9	43.9±6.6	42.6±6.5	42.7±4.5	44.9±7.3	**	A	C	D	B	E
N (%DM)	7.29±1.50	8.15±0.80	8.16±0.90	8.12±0.94	8.27±0.85	*	A	D	B	C	E
C:N ratio (weight base)	4.97±0.52	5.42±0.82	5.26±0.84	5.19±0.56	5.52±0.77	ns					
AFDM (%DM)	69.6±12.5	78.8±9.0	77.8±7.6	74.6±15.9	82.6±9.8	*	A	D	C	B	E
Energy (J mg^{-1} DM)	16.4±3.2	20.6±3.5	19.9±3.5	19.9±2.3	20.9±3.8	**	A	C	D	B	E

Based on Bray-Curtis similarity index by log-transformed data, zooplankton samples were separated into five groups (A-E, cf. Figure 4).
Values are mean±1sd. Between group differences was tested with one-way ANOVA and post-hoc test (Fisher's PLSD).
Any two groups not connected by underline are significantly different ($p<0.05$).
*: $p<0.05$, **: $p<0.01$, ***: $p<0.001$, ns: not significant, *n*: number of samples included.

Both 2003 and 2004, *N. cristatus* CV mass was greatest in the highest latitude stations and was decreased with decreasing latitudes. Compare within the same latitude, *N. cristatus* CV mass was consistently greater in 165°E than that in 165°W. Factor between *N. cristatus* CV mass in 165°E and 165°W (values in 165°E: 165°W) was varied between 1.2-13.6 (mean: 4.5).

Composition of *Neocalanus cristatus* CV to total zooplankton mass varied between 1.1% (39°00'N, 165°00'W in 2003) and 60.6% (48°00'N, 165°00'W in 2004) (Figure 6). Common to 2003 and 2004, composition of *N. cristatus* CV to total zooplankton mass was greater in the highest latitude stations and was decreased with decreasing latitudes. East-west differences in composition of *N. cristatus* to total zooplankton mass were not so evident (Figure 6).

Individual prosome length of *Neocalanus cristatus* CV varied between 5.8±0.3 mm (39°00'N, 165°00'W) and 6.9±0.3 mm (49°30'N, 165°00'E) (Figure 7). Both in 165°E and 165°W, prosome length of *N. cristatus* was larger in the high latitudes and decreased with decreasing latitudes. Carbon content of *N. cristatus* CV varied between 0.62 mg C ind^{-1} (36°00'N, 165°00'E in 2003) and 4.49 mg C ind^{-1} (49°30'N, 165°00'E in 2003) (Figure 7). In most of the stations, carbon content of *N. cristatus* individual was greater in the high latitude stations, and was greater in the 165°E within the same latitude, while it was not the case of southern end of 165°W in 2004 (42°30'N and 41°00'N).

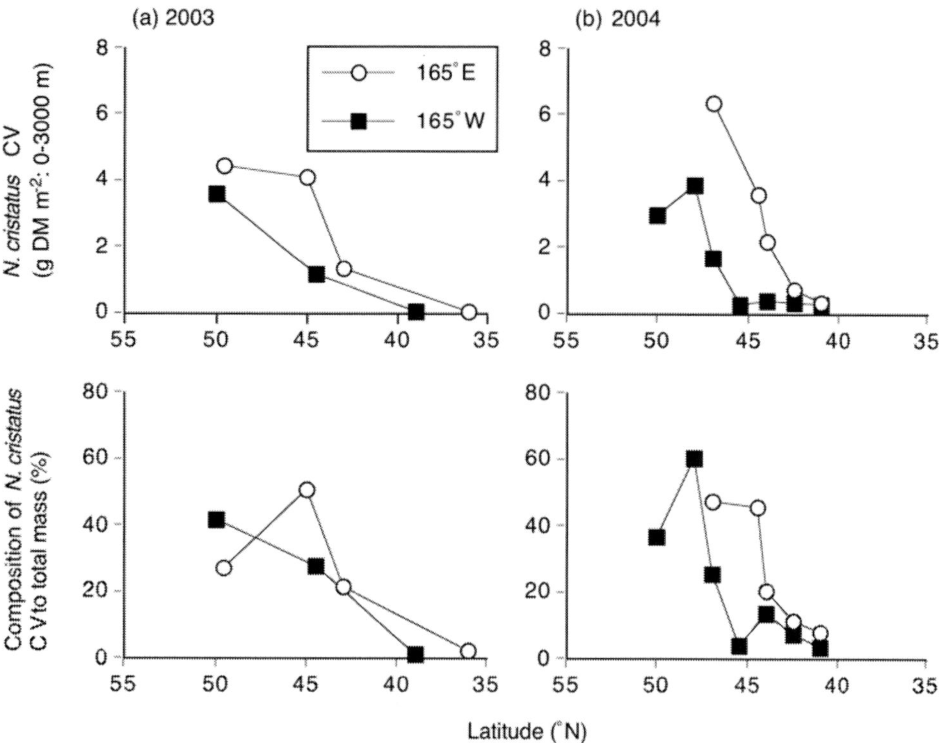

Figure 6. Latitudinal changes in biomass (dry mass) of copepod *Neocalanus cristatus* CV (upper) and their contribution (%) to total mesozooplankton mass (lower) during 2003 (a) and 2004 (b).

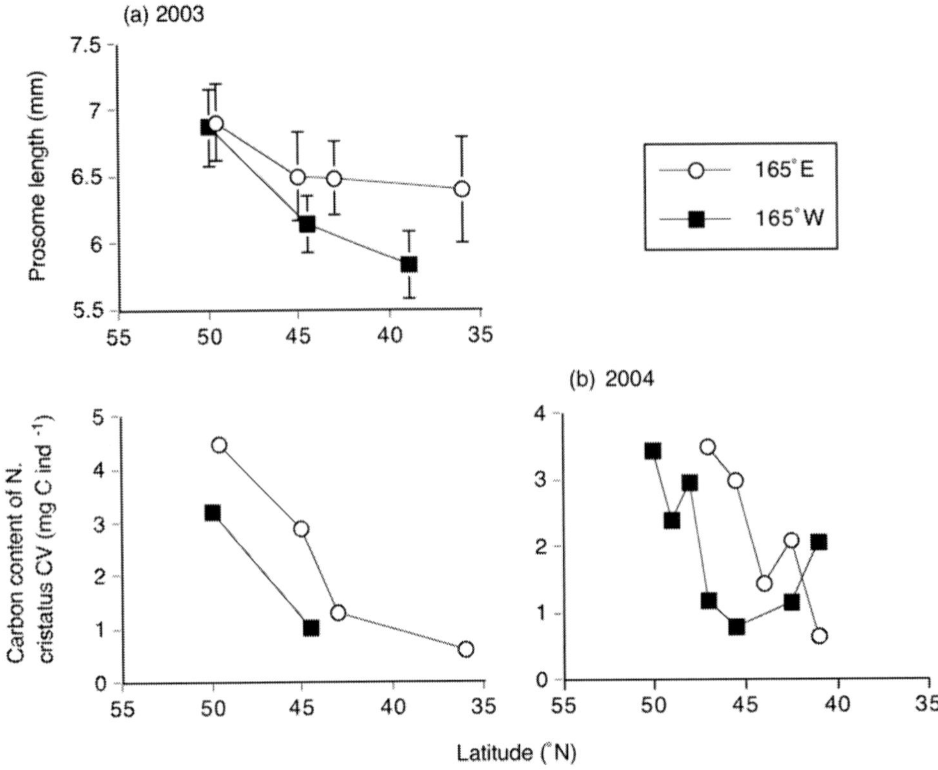

Figure 7. Latitudinal changes in prosome length (upper) and carbon content (lower) of *Neocalanus cristatus* CV during 2003 (a) and 2004 (b). Note that there are no data on prosome length in 2004. Vertical bars in prosome length indicate standard deviations.

Discussion

Total Zooplankton Mass

Total zooplankton mass observed in this study (5.9-28.0 g DM m^{-2}) is falls within values (1.3-27.4 g DM m^{-2}) reported from 0-5800 m of 44°N-25°N in the western North Pacific (Yamaguchi et al. 2005). In both 2003 and 2004, zooplankton masses were greater in the high latitude and were greater in the western (165°E) than those in the eastern (165°W) by a factor of 1.7 (Figure 3). These east-west differences in zooplankton masses are the first record in this region. What is the cause of these east-west differences? In terms of east-west differences in zooplankton ecology, life cycle pattern (presence of quiescence or not, growth and reproduction timing, and number of generation per year) of dominant copepods (*Neocalanus flemingeri*: Kobari and Ikeda 2001, *Eucalanus bungii*: Tsuda et al. 2004; Shoden et al. 2005, *Metridia pacifica*: Padmavati et al. 2004) and their body sizes (Tsuda et al. 2001, Kobari et al. 2003) are reported. Life cycle of copepods in the western subarctic Pacific is longer (*N. flemingeri*: Kobari and Ikeda 2001, *M. pacifica*: Padmavati et al. 2004) or shorter (*E. bungii*: Shoden et al. 2005) than those in the eastern Pacific. Longer generation length in the western region is attributed to the low temperature in the western region (Oyashio region) than that in the eastern region (Gulf of Alaska), while shorter generation length in the western region is considered to be related higher primary production in the western region than that in the eastern region. Body sizes of copepods are commonly larger in the western region than those in the eastern region, and are considered to be caused by the low temperature in the western region (Tsuda et al. 2001, Kobari et al. 2003). These large body sizes of dominant copepods in the western subarctic Pacific may be the cause of large zooplankton mass in the western region. For *Neocalanus* copepods, body size (prosome length) is varied regional by a factor of 1.22-1.25 (cf. Kobari and Ikeda 1999, 2001). If we assume masses varied with cubic body size, the difference in body size implies difference in mass by a factor of 1.8-1.9 (= 1.22^3 or 1.25^3). These values are close to the factor of zooplankton mass between western and eastern North Pacific observed in this study (1.7, Figure 3). In deed, factor between *Neocalanus cristatus* CV individual mass in 165°E and 165°W (values in 165°E: 165°W) was 4.5 observed (cf. Figure 7).

Chemical Contents

While the total zooplankton masses varied between east and west, their chemical components are similar between east and west. Chemical components of zooplankton showed latitudinal trends (group E in the north and group A in the south) in the upper epipelagic zone (0-50 m, Figure 5) and had vertical trends (from shallow to deep: group B<C<D, Figs. 4 and 5). The boundaries between group A and E in the upper epipelagic zone was corresponded to the subarctc front (SF) at 165°E in 2003 (Figure 2), while was at transition domain (region between SF and subarctic boundary [SB]) in the other transect/year (Figure 5). Since zooplankton community in the transition domain is known to have characteristics of subarctic fauna (cf. Kobari et al. 2003), the zooplankton communities of group A and E is considered to be those of subtropical and subarctic groups, respectively.

Interestingly, chemical contents of group A and E were very comparable, thus the group A was characterized by high water and low C, N, AFDM and energy, while the group E was characterized by low water and high C, N, AFDM and energy (Table 1). The differences in these chemical contents are considered to be caused by the differences in dominant zooplankton taxa (fauna). In the North Pacific, the subarctic fauna is characterized by the dominance of large grazing copepods which known to accumulate massive lipids in their body (= low water and high C and AFDM, cf. Omori 1970, Ikeda et al. 2004). While the subtropical fauna is characterized by the dominance of gelatinous pelagic tunicates (i.e. salps and doliolids) which contains high water and low C and AFDM (Bone 1998, Postel et al. 2000). These differences in zooplankton fauna between subarctic and subtropical region, may induced latitudinal gradients in chemical composition.

In terms of vertical changes, chemical contents of groups B, C and D are not so different, but masses (WM, DM, C, N, ash, and AFDM) and energy showed similar gradients across them, thus all in the order of B>C>D (Table 1). It indicates that the group separation of group B, C and D is mainly caused by the values of masses but not by the chemical contents.

Taxonomic Accounts

Both in 2003 and 2004, biomass of *Neocalanus cristatus* CV was consistently greater in the western than that in the eastern North Pacific (Figure 6). It suggests that this east-west gradient (greater in the western) is a common phenomenon continues year-round. Contribution of *N. cristatus* CV to total mesozooplankton mass was greater in the northern stations and reached 40-60%. Diapausing copepods (*Neocalanus* spp. and *Eucalanus bungii*) is known to contribute 70% of copepod mass in the subarctic Pacific (Yamaguchi et al. 2002a), they induce dominance (47-52%) of metazooplankton to whole plankton mass in this region (Yamaguchi et al. 2002b, 2004). In the present study, in 2004 when the sampling stations located with fine scale (every 45'N), region where the high *N. cristatus* contributed (>20% of total mass) was at north of 46°N at 165°W, while was at north of 43°N at 165°E (Figure 6). Thus, the high *N. cristatus* region located more north in 165°W than that in 165°E, it may related to the subarctic front (SF) in 165°W was located more north than that in 165°E (Figure 2).

Body size of copepods is known to have negative relationship between their habitat temperatures (cf. Mauchline 1998). In the present study, prosome length of *Neocalanus cristatus* CV was larger in the northern stations and compared within the same latitude, it was larger in the western (165°E) region (Figure 7). It may imply large carbon content per individual in the northern station and western (165°E) stations (Figure 7). In 2004, there are no data on prosome length, while there are similar patterns (larger in the northern and western stations) except southern two stations in 165°W.

CONCLUSION

In the present study, total zooplankton masses were consistently greater in the western (165°E) North Pacific (Figure 3). It may caused by the dominance of large sized copepod *Neocalanus cristatus* CV in the western Pacific (Figure 6) and together with their large body size (which implies greater masses) in the western region (Figure 7). This large body size of *N. cristatus* CV may related to the low temperature condition in the western North Pacific (Figure 2). In the last decade, east-west differences in plankton community of the North Pacific are reported. Thus, phytoplankton density and primary production are known to be higher in the western than those in the eastern North Pacific (Shiomoto and Asami 1999, Shiomoto and Hashimoto 2000). This east-west difference is considered to be caused by the differences in concentration of dissolved-iron (higher in the western) (Harrison et al. 1999, Suzuki et al. 2002). East-west differences in phytoplankton community effects zooplankton abundance, biomass, and community in these region (Mackas and Tsuda 1999), and higher transfer efficiency in the western region is suggested (Taniguchi 1999). The consistently greater mesozooplankton masses in the western North Pacific in this study indicate these east-west differences is a common phenomenon throughout the lower trophic level of the North Pacific Ocean.

ACKNOWLEDGMENTS

We are grateful to T. Ikeda for valuable comments on earlier version of manuscript. We thank the captains and crews of the T.S. *Oshoro-Maru* for their cooperation during the field sampling. We also thank F. Imao, F. Sano, S. Tachibana and N. Yamada for their help during the field sampling. This study is made with JSPS Grant-in-Aid for Scientific Research (S) 16108002 (Historical transition and prediction of Northern Pacific ecosystem associated with human impact and climate change).

REFERENCES

Bone, Q. (1998). *The Biology of Pelagic Tunicates*. Oxford University Press, Oxford, 362 pp.

Bray, J. B., and Curtis, J. T. (1957). An ordination of the upland forest communities of southern Wisconsin. *Ecol. Monogr., 27*, 325-349.

Favorite, F., Dodimead, J. A., and Nasu, K. (1976). Oceanography of the subarctic Pacific region, 1960-1971. *Bull. Int. North Pacific Fish. Comm., 33*, 1-87.

Gnaiger, E. (1983). Calculation of energetic and biochemical equivalents of respiratory oxygen consumption. In Gnaiger, E. and Forstner, H. (Eds.), *Polarographic Oxygen Sensors* (pp. 337-345). Springer, Berlin.

Gnaiger, E., and Shick, J. M. (1985). A methodological note on CHN-stoichiometric analysis. *Cyclobiosis Newsletter, 2,* 2 -4.

Harrison, P. J., Boyd, P. W., Varela, D. E., Takeda, S., Shiomoto, A., and Odate, T. (1999). Comparison of factors controlling phytoplankton productivity in the NE and NW subarctic Pacific gyres. *Prog. Oceanogr., 43*, 205-234.

Hernández-León, S. and Ikeda, T. (2005). A global assessment of mesozooplankton respiration in the ocean. *J. Plankton Res., 27*, 153-158.

Ikeda, T., Sano, F., and Yamaguchi, A. (2004). Metabolism and body composition of a copepod (*Neocalanus cristatus*: Crustacea) from the bathypelagic zone of the Oyashio region, western subarctic Pacific. *Mar. Biol., 145*, 1181-1190.

Kobari, T., and Ikeda, T. (1999). Vertical distribution, population structure and life cycle of *Neocalanus cristatus* (Crustacea: Copepoda) in the Oyashio region, with notes on its regional variations. *Mar. Biol., 134*, 683-696.

Kobari, T., and Ikeda, T. (2001). Life cycle of *Neocalanus flemingeri* (Crustacea: Copepoda) in the Oyashio region, western subarctic Pacific, with notes on its regional variations. *Mar. Ecol. Prog. Ser., 209*, 243-255.

Kobari, T., Tadokoro, K., Shiomoto, A., and Nishimoto, S. (2003). Geographical variations in prosome length and body weight of *Neocalanus* copepods in the North Pacific. *J. Oceanogr., 59*, 3-10.

Mackas, D. L., and Tsuda, A. (1999). Mesozooplankton in the eastern and western subarctic Pacific: community structure, seasonal life histories, and interannual variability. *Prog. Oceanogr., 43*, 335-363.

Mauchline, J. (1998). The biology of calanoid copepods. *Adv. Mar. Biol., 33*, 1-710.

Michaels, A. F., and Silver, M.W. (1988). Primary production, sinking fluxes and the microbial food web. *Deep-Sea Res., 35A*, 473-490.

Motoda, S. (1959). Devices of simple plankton apparatus. *Mem. Fac. Fish. Hokkaido Univ., 7*, 73-94.

Omori, M. (1970). Variations of length, weight, respiratory rate, and chemical composition of *Calanus cristatus* in relation to its food and feeding. In Steele, J. H. (Ed.) *Marine Food Chains*. (pp. 113-126). Oliver and Boyd, Edinburgh.

Padmavati, G., Ikeda, T., and Yamaguchi, A. (2004). Life cycle, population structure and vertical distribution of *Metridia* spp. (Copepoda: Calanoida) in the Oyashio region (NW Pacific Ocean). *Mar. Ecol. Prog. Ser., 270*, 181-198.

Postel, L., Fock, H., and Hagen, W. (2000). Biomass and abundance. In: Harris, R. P., Wiebe, P. H., Lenz, J., Skjoldal, H. R., and Huntley, M. (Eds.), *ICES Zooplankton Methodology Manual* (pp. 83-192). Academic Press, San Diego.

Shiomoto, A., and Asami, H. (1999). High-west and low-east distribution patterns of chlorophyll a, primary productivity and diatoms in the subarctic North Pacific surface waters, midwater 1996. *J. Oceanogr., 55,* 493-503.

Shiomoto, A., and Hashimoto, S. (2000). Comparison of east and west chlorophyll a standing stock and oceanic habitat along the Transition domain of the North Pacific. *J. Plankton Res., 22*, 1-14.

Sheldon, R. W., Sutcliffe Jr., W. H., and Paranjape, M. (1977). Structure of pelagic food chain and relationship between plankton and fish production. *J. Fish. Res. Bd. Can., 34*, 2344-2353.

Shoden, S., Ikeda, T. and Yamaguchi, A. (2005). Vertical distribution, population structure and life cycle of *Eucalanus bungii* (Copepoda: Calanoida) in the Oyashio region, with notes on its regional variations. *Mar. Biol., 146,* 497-511.

Suzuki, K., Liu, H., Saino, T., Obata, H., Takano, M., Okamura, K., Sohrin, Y., and Fujishima, Y. (2002). East-west gradients in the photosynthetic potential of phytoplankton and iron concentration in the subarctic Pacific Ocean during early summer. *Limnol. Oceanogr., 47*, 1581-1594.

Taniguchi, A. (1999). Differences in the structure of the lower trophic levels of pelagic ecosystems in the eastern and western subarctic Pacific. *Prog. Oceanogr., 43*, 289-315.

Terazaki, M., and Tomatsu, C. (1997). A vertical multiple opening and closing plankton sampler. *J. Adv. Mar. Sci. Tec. Soc., 3*, 127-132.

Tsuda, A., Saito, H., and Kasai, H., (2001). Geographical variation of body size of *Neocalanus cristatus*, *N. plumchrus* and *N. flemingeri* in the subarctic Pacific and its marginal seas: implications for the origin of large form of *N. flemingeri* in the Oyashio region. *J. Oceanogr., 57*, 341-352.

Tsuda, A., Saito, H., and Kasai, H., (2004). Life histories of *Eucalanus bungii* and *Neocalanus cristatus* (Copepoda: Calanoida) in the western subarctic Pacific Ocean. *Fish. Oceanogr., 13*, S10-S20.

van der Meeren, T. and Næss, T., 1993. How does cod (*Gadus morhua*) cope with variability in feeding conditions during early larval stages? *Mar. Biol., 116*, 637-647.

Yamaguchi, A., Watanabe, Y., Ishida, H., Harimoto, T., Furusawa, K., Suzuki, S., Ishizaka, J., Ikeda, T. and Takahashi, M. M. (2002a). Community and trophic structures of pelagic copepods down to the greater depths in the western subarctic Pacific (WEST-COSMIC). *Deep-Sea Res. I, 49*, 1007-1025.

Yamaguchi, A., Watanabe, Y., Ishida, H., Harimoto, T., Furusawa, K., Suzuki, S., Ishizaka, J., Ikeda, T. and Takahashi, M. M. (2002b). Structure and size distribution of plankton communities down to the greater depths in the western North Pacific Ocean. *Deep-Sea Res. II, 49*, 5513-5529.

Yamaguchi, A., Watanabe, Y., Ishida, H., Harimoto, T., Furusawa, K., Suzuki, S., Ishizaka, J., Ikeda, T. and Takahashi, M. M. (2004). Latitudinal differences in the planktonic biomass and community structure down to the greater depths in the western North Pacific. *J. Oceanogr., 60*, 773-787.

Yamaguchi, A., Watanabe, Y., Ishida, H., Harimoto, Maeda, M., Ishizaka, J., Ikeda, T. and Takahashi, M. M. (2005) Biomass and chemical composition of net-plankton down to greater depths (0-5800 m) in the western North Pacific Ocean. *Deep-Sea Res. I, 52*, 341-353.

In: The Pacific and Arctic Oceans
Editor: Kallen B. Tewles

ISBN: 978-1-60692-010-7

Chapter 4

SELF-ORGANIZED DYNAMICS OF THE ARCTIC SEA-ICE COVER

Alexandre Chmel[1] and Victor Smirnov[2]
[1]Fracture Physics Department, Ioffe Physico-Technical Institute, Russian Academy of Sciences, 194021 St. Petersburg, Russia
[2]Physics of Ice Laboratory, Arctic and Antarctic Research Institute, 38 Bering str., 199397 St. Petersburg, Russia

ABSTRACT

The mechanical behavior (fragmentation and drift) of the sea-ice in the Arctic Ocean is considered in the light of its space and time invariance concluded from the analysis of the geometrical pattern of fragmented sea-ice cover and drift dynamics, respectively. The well-pronounced fractal properties of the Arctic sea-ice cover (ASIC) are interpreted in terms of the concept of self-organized criticality, which implies the perpetual balancing of the open statistical system on the border of stability with multiple cycles of fracturing and restoring occurring at any time and throughout the system. The permanent criticality of this kind needs a dynamic long-range and long-term connectedness of the system, which in the case of ASIC is realized and maintained through a variety of wave and oscillation processes that promote the energy transfer and exchange in the sea-ice cover. An analogy between the energy exchange in the sea-ice drift and the energy release in unstable tectonic formations (earthquakes) is drawn and discussed. The sources of the information were the databases of field observations carried out at the Russian ice-research stations "North Pole", supplemented with the remote technique data available from other origins or reported by other research groups.

1. INTRODUCTION

The Arctic sea-ice cover (ASIC) is the open, non-equilibrium, dynamic system. Its behavior is governed by the mechanics of rigid body evolving under conditions inherent to the geophysical object of climate scale. The sea-ice cover interacts with the ocean and atmosphere whose highly changeable forcing and thermophysical effects determine the variable properties of the ASIC.

Two fundamental (and interrelated) processes are characteristic for the ASIC: the permanent drift of the sea-ice and cycles of fracturing and refreezing. The mobility of ice sheets causes continuous redistribution of areas of open water and reflectivity of ice-cover with corresponding effects on both regional and global climatic characteristics through changes in the temperature and albedo. In addition, the sea-ice drift and freezing-fragmentation cycles are of importance for some kinds of the human activity in the cold regions, such as the navigation and the functioning of marine engineering structures.

The intrinsic mobility of the ASIC is a sum of quasi-stationary drift (due to, mainly, the wind forcing), and local oscillations of ice sheets (due to their shearing and collisions). The impact interactions excite elastic waves, which propagate over vast areas of the sea-ice cover thus performing the dynamic connectedness of the ASIC. The long-range dynamic continuity is supplemented with the long-term compression-stimulated oscillations, which induce some memory effects in the drift dynamics.

The most recent studies performed in the last decade have shown that the typical mechanic characteristics of the sea-ice dynamics, such as the deformation rate (Marsan et al. 2004) and the force interactions (Chmel et al., 2007), exhibit features of self-organization and self-similarity. These features are particular representations of the fractal properties of the ASIC, which manifest themselves in the geometry and numerical distribution of components in the collection of sea-ice pieces (Weiss and Marsan, 2004) and in the time invariance of the ice pack drift (Chmel and Smirnov, 2007). The problem of self-organization in the dynamic systems is highly debated in geophysics in relation to, mainly, seismic activity (Bak and Tang, 1989; Sornette and Sornette, 1989; Cristensen and Olami, 1992; Bak et al. 2002); the revealed in the end of the last century scaling properties of the main characteristics of the tectonic processes stimulated the appearance of a new concept of the earthquake prediction based on the consideration of Earth's crust as the self-organizing critical system.

In this chapter we shall apply the developed in seismology and fracture physics methods of statistical analysis of open self-organizing systems to the description of the ASIC as the non-equilibrium, muli-component system. Various aspects of the self-organization of the sea-ice dynamics will be considered using the data of field observations carried out at the Russian ice-research stations "North Pole" (NP) established on the Arctic ice pack in different years, including the data of a case study of the self-organized dynamics that was carried out in the station NP 32 (2003 to 2004). To provide an extended view of the problem, the relevant results of other research groups that developed the same scenario utilizing the remote sensing technique will be reviewed and discussed.

2. Self-Similarity and Self-Organization

The sea-ice dynamics is closely related with events of fracture, which result in a complicated pattern of ridges, leads and cracks in the ASIC. From the viewpoint of the convention mechanics, the fragmentation and lead formation in the sea-ice cover are specific natural phenomena, which cannot be regarded in common terms with, respectively, the fracture events and faulting in "truly" *geo*physical objects (tectonic formations) of the same scale level. The apparent difference the elements of the Earth's crust emerges from three main inconsistencies: i) the mechanics of rigid body implies a three-dimension scenario of the deformation and fracturing, while the ASIC is a kind of a thin film lying on the liquid substrate; ii) the fracture and faulting in the Earth's crust are, in fact, irreversible in contrast to the ASIC fragmentation, which permanently alternates with the freezing (healing) cycles (Korsnes et al.); iii) the crustal deformation is not turbulent, while the deformation of the sea-ice exhibits significant local irregularities (Weiss and Marsan, 2004). At the same time, the application of the methods of the modern statistical physics to both these objects of geophysical scale performed in the last decade by geophysicists studying the non-linear processes has shown that despite the great difference between the physical properties of the lithosphere and cryosphere one could point out a feature that establishes a surprising closeness between these media.

This common feature is the self-similarity and the self-organization of the dynamics.

What does mean the statistical self-similarity? Let an observer looks at the satellite image of a giant lead, such as shown in Figure 1. The lead consists of multiple smaller fragments; let us consider each "straight" portion of the lead (that is portions between neighboring points of turning, branching, or termination) as an individual crack. Obviously, the longer crack, the smaller amount of such cracks. Let the function N(r) gives one the dependence of the number of cracks, N, on their length, r.

The function of this kind would characterize the spatial correlation between events of crack formation (because of comprehensible physical reasons, the appearance of a primary crack would affect the nucleation of other cracks in its vicinity). A lot of processes in the nature exhibit the exponential decay of the correlation between discrete events both in space and time. However, the fracture as well as many other critical processes exhibits another behavior. The criticality implies the cooperativity, that is the long-range correlation. Numerous experiments on cracking and fragmentation of solids (including the ice breakage, which we shall consider in the next Section) show that the number of "damaged sites", N, decreases with their size following the power law:

$$N(r) \propto r^{-D} \tag{1}$$

where D is the fractal dimension. The statistics that follows the power law indicates the lack of characteristic sizes since the function N(r) satisfies the equation of the scale invariance:

$$N(\lambda r) = \lambda^{-D} N(r), \tag{2}$$

where λ is the scale factor. In our example, this means that the total number of straight portions of the branching lead of different sizes would change with the scale of observation λr as given by the Eq. (2). In simple phrase, when an observer is looking at the object like that that shown in Figure 1 using different magnifications (for example, from different distances), he or her would see every time a new pattern of straight portions of the lead. This pattern depends only on the λ but not on the resolution scale r. In the case of a truly random lead path, the observer would see a smoothed line when looking from the large distance: the details disappear with the scale increase. A similar approach one could realize when looking, for example, at the trajectory of an iceberg motion (Korsnes, 1991).

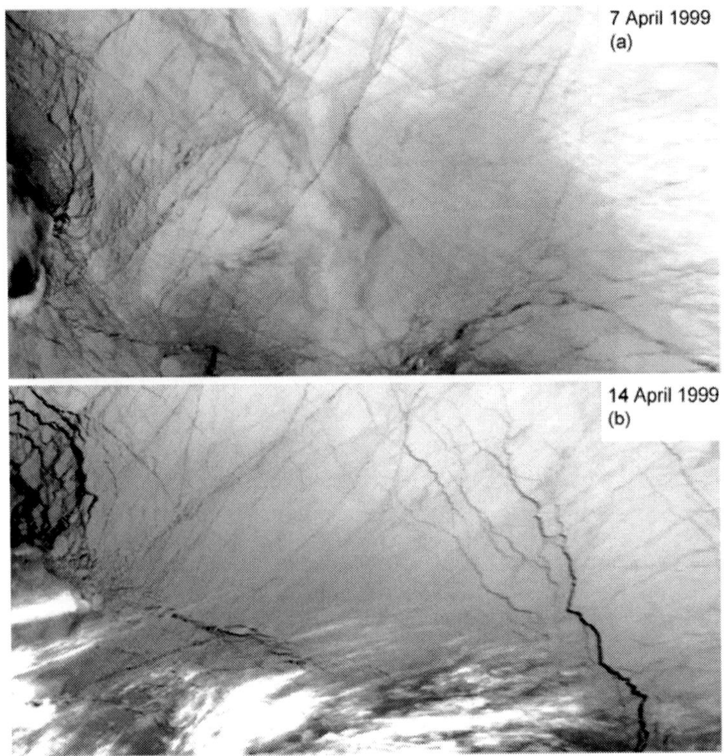

Figure 1. The formation of a large lead imaged from the NOAA 16 satellite in the Arctic Ocean on 7 April and 14 April 1999. The newly formed lead was centered around, approximately, N84°; E110°, and propagated over almost 10^2 km.

The objects that do not exhibit characteristic length scale (that is those that follow the power law in forms (1) or (2)) are called fractal objects. The fracturing solid is the case. The power law distribution of fracture products results from the long-range correlation of the fracture process itself: each new breaking of structural links affects not only the neighboring sites but the whole ensemble of sites in a way to maintain the critical state (Caldarelli et al., 1996). The lack of any characteristic length or time (spatial or temporal fractality) is typical for critical dissipative structures, such as a solid in the state of critical phase transition of the second kind. In this sense, the fracturing is an example of dynamic phase transition (Bauchaud, 1997). However, there is an essential difference between the classical phase transition and a marginal critical state of the dynamic system driven by the outer forcing: the latter one approaches its marginal state without necessity of fine tuning of some external parameters. There are many processes in the nature when the character of the dynamics changes from a fully random process to self-similar one (or from spatially or temporarily restricted self-similarity to more expanded self-similarity). The dynamics of this kind is called "self-organized". The fracture is the example of the self-organized process.

The latter statement can be illustrated by the experiment on ice fracturing carried on by Weiss and Gay (1998). They performed a series of compression tests of fresh-water ice under the room conditions. The experiments were interrupted before general failure. Fracture patterns on transversal sections of loaded samples were analyzed under a microscope. The measured length distributions of damage sites at successive stages of loading allowed authors to assess the alteration of the scaling properties of damaged ice. At the initial stage of loading the fracture pattern, which consisted of isolated cracks, exhibited the power law distribution over one order of magnitude. As the deformation grown, both the damage clusters and isolated fragments of material appeared. Simultaneously, the scale invariance of fracturing spreaded over a larger scale range (for account of covering smaller scales). This means that this dynamic system evolves with expanding the spatial range of the self-similar organization. However the most striking result was obtained when analyzing the distribution of fragment sizes after failure of the sample. A power law scaling was observed over the entire scale range accessible for measurements. At the same time, no dependence of the power exponent on the loading conditions (creep; constant strain rate; constant stress-rate) was detected; consequently, the fracturing system self-organizes spontaneously to a self-similar collection of damaged sites whose crucial scaling parameter (scaling exponent) is characteristic for the given critical state of the system itself but not for the external forcing.

3. Spatial Scaling of Fragmented Sea-Ice

The simplest method to demonstrate the spatial invariance of the sea-ice fragmentation is to construct the cumulative size distribution of ice pieces N(>L). The function N(>L) specifies the number of ice-floes whose size (e. g. the largest linear dimension) exceeds a characteristic length L available for measuring in the images made from a plane or space. An example of the analysis of this kind was reported by Weiss and Marsan (2004). A SPOT satellite image of fragmented sea-ice cover in the geographical point NW of Queen Elizabeth Islands (Figure 2) was processed in order to verify the scaling character of the fragmentation. The result is shown in Figure 3.

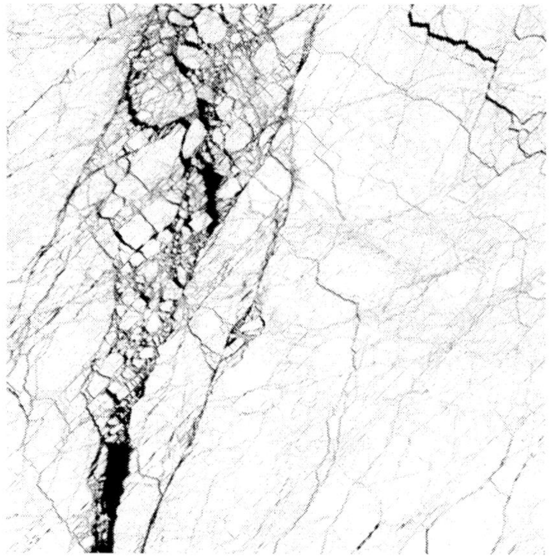

Figure 2. Digitalized SPOT satellite image of the sea-ice cover taken on 6 April 1996, centered around N80° 11′, W108°33′ and covering 59 × 59 km^2. Reproduced with permission from (Weiss and Marsan, 2004).

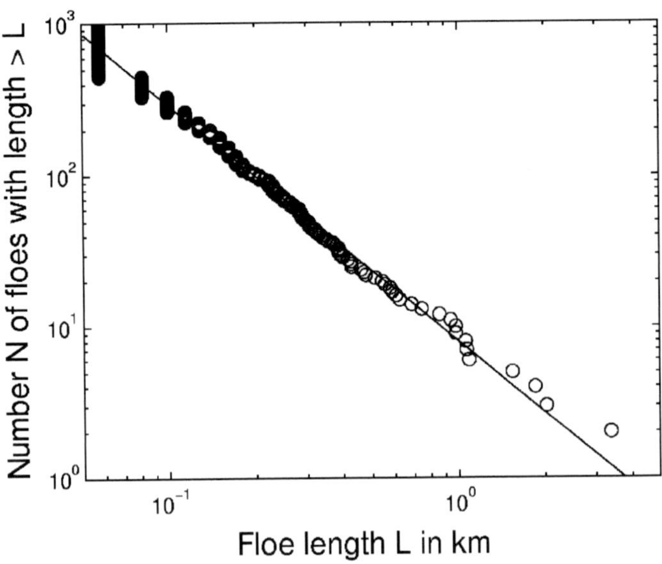

Figure 3. Cumulative distribution of floes sizes, N (> L), calculated from the analysis of the satellite image shown in Figure 2. Adapted with permission from (Weiss and Marsan, 2004).

One can see that the distribution function N(>L) exhibits a log-linear dependence, that is follows the power law $N(>L) \propto L^{-\beta}$.

The scaling geometry of the crack-and-lead pattern was also demonstrated by Chmel et al. (2005) who analysed the satellite images with the help of the technique developed by Mandelbrot et al. (1984) for determining the fractal dimension of fractured surface. Figure 4 shows an image of the Arctic region north from Spitsbergen and up to approximately N86°.

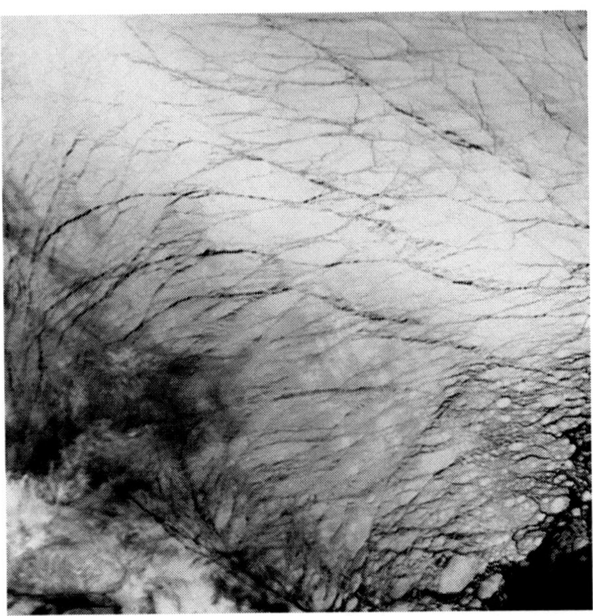

Figure 4. A NOAA-16 satellite image of the region north from Spitsbergen (made on 10 December 2003).

The area (S) and perimeter (L) of ice-floe fragments limited by cracks and open water were determined using a special computer program adapted for this task. According to Mandelbrot (2006), in the case of the fractal geometry of two-dimensional objects, the relation between the area of fragment's surface and the length of its "coastline" would satisfy the power law:

$$S^D \propto L^2 \quad (3)$$

A set of accounted ice pieces of different sizes was composed of ice fields situated throughout the available area. The measured values of the area and perimeter were plotted in doubly logarithmic coordinates (Figure 5). Again, as like as in the case of cumulative size distribution (Figure 3), the points fall on straight graph, being the proportionality between log L and log S extended over three orders of magbitude. The power law (3) evidences the spatial invariance of the fragmented sea-ice cover.

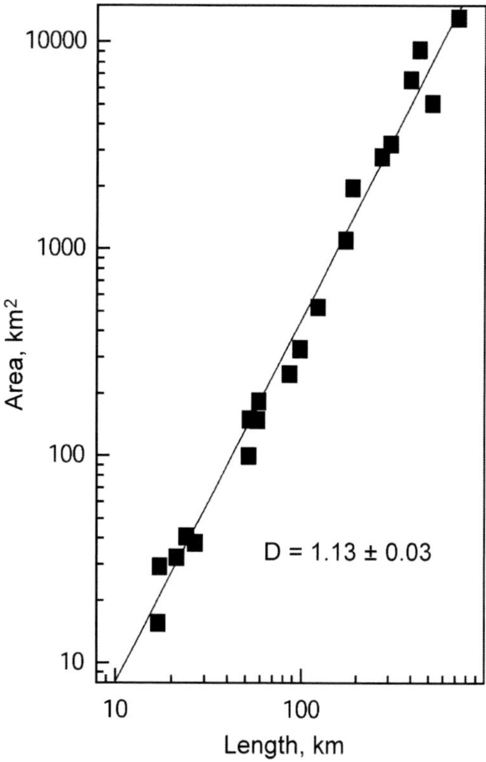

Figure 5. Area versus perimeter dependences for ice-floes resolved in the image shown in Figure 4. The straight line fits the power law dependence. The accuracy of the determination of D was estimated from the deviation of the slope at the end points of log-log plots with respect to the best-fit data of S versus L dependences.

The fractal dimension value can be determined from the graph's slope using Eq. (3). In our case, D = 1.13 ± 0.03. The fractal dimension, which characterizes the dimension of the systems's dynamic attractor, is not a universal parameter of the fragmented sea-ice. Figure 6 shows the results of the area-versus-perimeter measurements carried out for the portions of the sea-ice cover depicted in Figure 1.

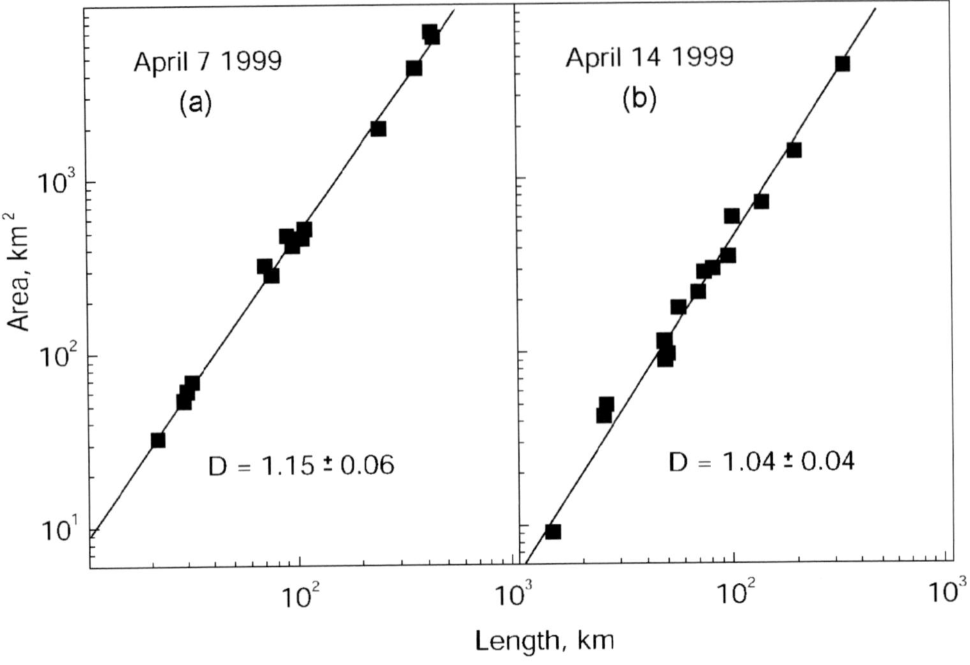

Figure 6. Area versus perimeter dependences for ice-floes resolved in images shown in Figure 1. The shown values of D were found from the slope of straight lines fitted to S versus L dependences prior to the large lead formation (a), and after this event (b).

One can see that the fractal dimension of the crack-and-lead pattern decreased from 1.15 ± 0.06 down to 1.04 ± 0.04 after a cycle of sea-ice fragmentation. In terms of the fractal geometry, the propagating leads and cracks destroy two-dimensional formations with enhancing one-dimensional patterning characterized by the smaller D-value. More detailed analysis of the response of the fractal dimension to the events of large-scale fragmentation revealed that the value $D \approx 1.14$ is typical for "quiet" periods of the sea-ice drift, while the important events of fracturing cause its decrease with restoring the initial value in a few weeks after the fragmentation (Chmel et al., 2007).

The fractal pattern of the fragmented ice pack results from the widely extended dynamics of ASIC components. To produce some organized (i. e. space-invariant) structures, the dynamics should be also organized in space, that is to be long-range correlated. In addition, the sea-ice motion should manifest some "memory" effects (temporal ordering), since otherwise, the "instantaneous" configurations of a collection of ice pieces would not form sequences going to the attractor of the systems's dynamics (scale-invariant pattern). In other words, one should expect that the dynamics of sea-ice motion would be self-organized in space and time.

4. SEA-ICE DRIFT DYNAMICS

The spatiotemporal invariance is typical for critical systems. In the conventional physics, the criticality is usually related to the finely tuned states, such as phase transitions of the second kind, which are characterized by some accurately defined thermodynamic parameters. In the case of non-equilibrium system, the critical state (in the sense of infinite space-time correlation) could be reached at various combinations of external parameters due to dynamic interaction of the system with the energy flow that stimulates its evolution.

As a rule, the spatial scaling and the temporal scaling in the non-equilibrium critical systems supplement each other (Maslov et al., 1994). Aksenov (1999) reported the interrelation between the fractal spatial pattern and temporal structure of ice deformation, which, as it had been concluded from the field experiments, exhibits the scaling properties. The scaling of sea-ice cover deformation rate on the area up to 10^3 km was reported by Marsan et al. (2004) who used in their investigation the satellite radar data. These experimental evidences of correlated deformation in the ASIC are of great importance because the occurrence of large-scale mechanical perturbations is directly related with the non-uniform stress distributions in drifting sea-ice. The long-range deformations stimulate co-operative fracturing. Otherwise, the ice-cracking would be highly localized.

The temporal redistribution of stresses and deformations in the dynamic system of ASIC is connected with the irregular accelerations of different parts of ice cover. The drift of massive ice fields is caused mainly by quasi-stationary surface wind shear stress (Marsan et al., 2004) and, to a lesser extent by ocean currents (Weiss, 2003), and affected by the action of random factors emerged from accelerations due to fluctuations in the atmospheric pressure and ambient temperature (Lewis et al., 1994), short time atmospheric turbulence and shore interaction. The high-amplitude local accelerations are caused by the mechanical interaction between ice fields (Martin and Drucer, 1991; Smirnov, 1996).

These random accelerations whose amplitudes vary significantly over the ice cover is the main source of short stress pulses. The duration of impact-induced series is 10^2 – 10^3 s (Smirnov, 1996) ; the stress in spikes reaches 400 kPa (Tucker and Perovich, 1992; Smirnov, 1996). Tucker and Perovich (1992) called the impact-induced stresses "dynamic" in contrast to "thermal" stresses caused by non-uniform changes in the ice temperature.

Thus, the fracture process must reflect, in general, the principal trends of the ice drift irregularity. Therefore, the above-mentioned spatial scaling of the collection of ice floes is closely related with the temporal fractality of the sea ice dynamics.

Hereafter, we shall demonstrate various statistical representations of the temporal organization of the drift-related events occurring in the ice pack, and discuss the potential mechanism of the long-range dynamic interactions, which determine the space-time fractal structure of the ASIC.

4.1. Temporal Correlation

The dynamics of the local sea-ice motion was monitored using GPS transducers established on the ice fields. In order to determine the accuracy of positioning, and to derive the reliable information against a background of both natural and man-caused noise, a special experiment was carried out. Two identical GPS transducers were actuated simultaneously on the same ice-field at the distance of 180 m. Then, without interrupting the measurements, one of the transducers was displaced for 9 m, and the positioning was continued. The result is shown in Figure 7.

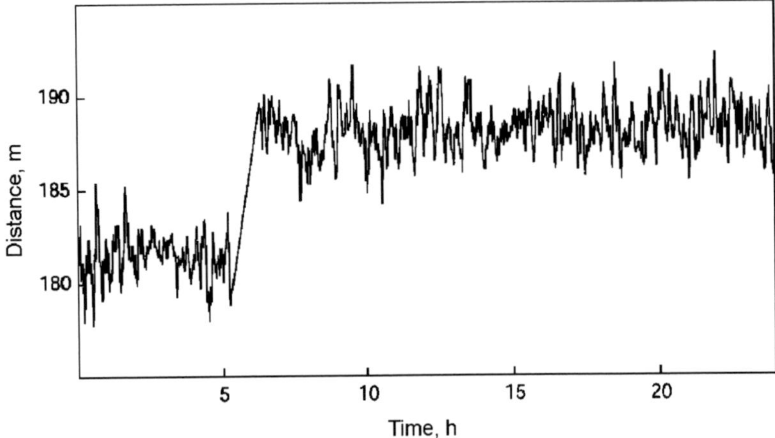

Figure 7. Distance between two GPS transducers as calculated from readings of both units. A "jump" resulted from a displacement of one of them for 9 m.

One can see that this displacement was reflected by the measuring system with the accuracy about ± 3 m with sporadic spikes up to 5 m (in more prolonged series some spikes due to rare artefacts reached 15 m). Under the assumption that the distance between the transducers does not vary till no breaks of the ice continuity take place, and taking into account that the dynamic and thermal deformations of the ice-field were significantly smaller than the scale of measurements, one could conclude that the observed noise fluctuations were predominantly due to the positioning errors in the data of two independent GPS channels.

On the base of the GPS data, the ice-field speed (V) and accelerations $|A| = |\Delta V|/\Delta t$ were calculated (here ΔV is the variation of drift rate measured in regular intervals $\Delta t = 10$ min).

In order to reveal the temporal invariance of the force interactions in the fragmented ice pack (given by local accelerations of an ice-field), Chmel et al. (2005) computed the correlation integral introduced by Grassberger and Procaccia (1983):

$$C(t) = [2/N(N - 1)]\Sigma_{i,j} H(t - t_{i,j}) \tag{4}$$

for a succession of accelerations recorded in the research station NP 32 in the period of time from 30 August 2003 to 3 October 2003 (Figure 8). Here C(t) is the probability for two events to be separated in time by less than t; N is the total number of "events" (that is all accelerations of any amplitudes); t is the time scale, H is the Heaviside function (H = 1 if the interval between events is less than t; otherwise, H = 0). If the function C(t) exhibits a power law, the succession of events can be regarded as temporal fractal. In addition, the definition (4) allows one to assess the relative roles of accelerations of different amplitudes (induced, probably, by different sources) in the time-invariant component of the correlated sea-ice drift process.

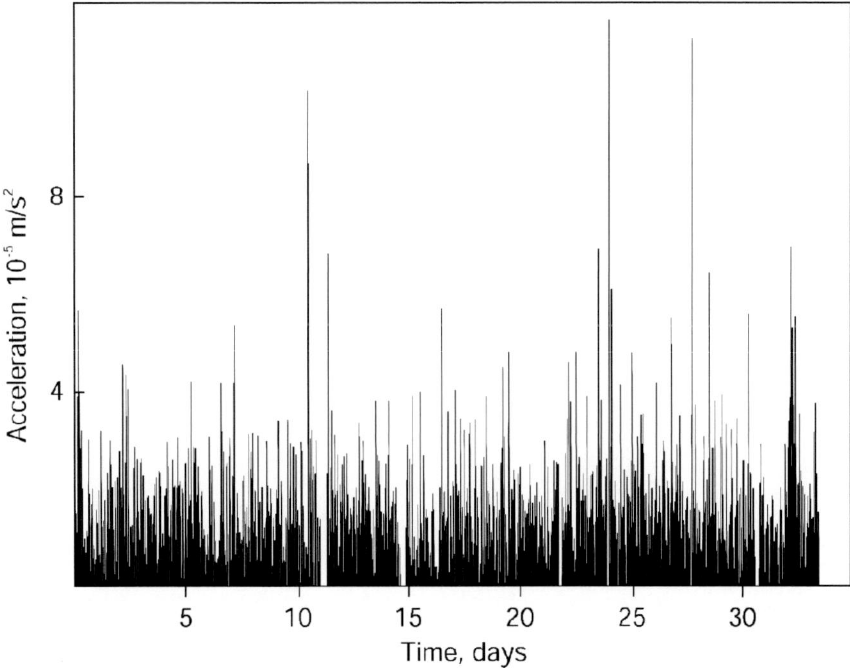

Figure 8. Amplitudes of accelerations measured in the period of time from 30 August 2003 to 3 October 2003. The accelerations are referred to 10 min changes of the ice field velocity. From the database of NP 32.

For that, one should restrict himself or herself by taking into account only accelerations of particular amplitude, for example, only those that satisfy the inequality $A > A_{lim}$. In this study, the C(t) was calculated, first of all, for the accelerations whose amplitudes were greater than $A_{lim} = 5\times10^{-6}$ m/s^2. In other words, the "acceleration events" were defined as accelerations that satisfy the condition $A > 5\times10^{-6}$ m/s^2. This condition covered 40 per cent of all the events detected.

The results are shown in Figure 9a. One can see that the plot gives $C(t) \sim t^{0.97}$ what is very close to a linear dependence characteristic for the random Poisson process. This means an almost full absence of the time correlation.

Figure 9b shows the C(t) obtained under condition $A > 13\times10^{-6}$ m/s^2. Only 4 per cent of detected events satisfy this condition while 96 per cent of low-amplitude accelerations were ignored. The straight-line portion of the graphic gives a power dependence $C(t) \sim t^{0.85}$ This is clearly below a linear scaling $C(t) \sim t$. Consequently, the accelerations with amplitudes $A > 13\times10^{-6}$ m/s^2 are correlated in time.

A more detailed information on the relative significance of accelerations of various amplitudes in the temporal invariance of sea-ice drift can be obtained from the analysis of the time separations ("waiting times") between accelerations of different importance. For this purpose, the same time sequence of values A that utilized for computing the Grassberger-Procaccia integral (4) was used to construct the distribution of waiting times with respect to the amplitudes of "expecting" accelerations, $\tau(>A)$. In other words, τ is the minimal time separation between recorded events of acceleration higher than A.

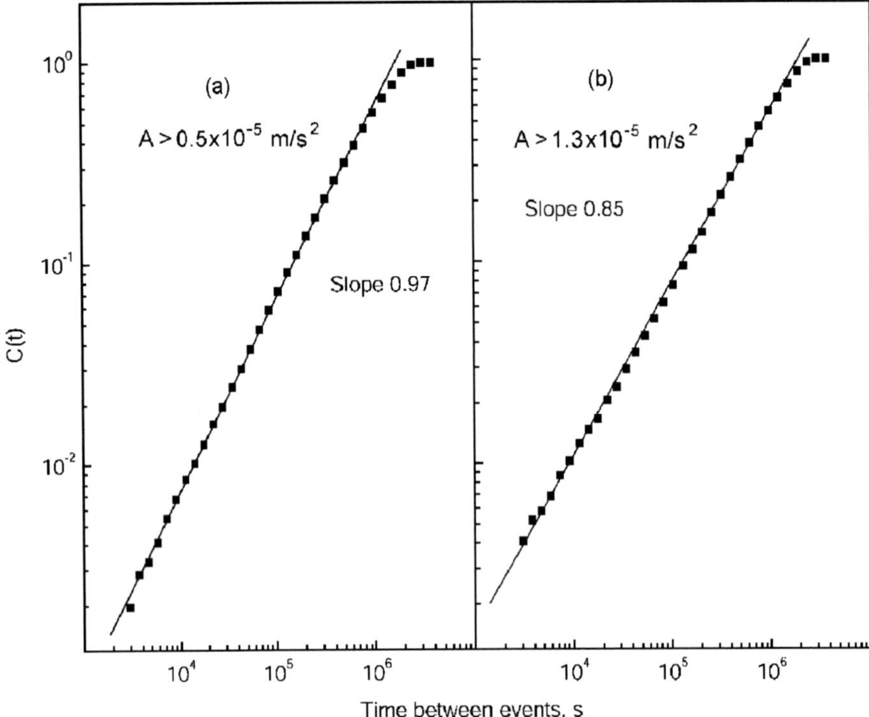

Figure 9. Correlation integral for accelerations satisfying different conditions $A > A_{lim}$. The straight lines fit the power law $C(t) \sim t^{0.97}$ (a) and $\sim t^{0.85}$ (b).

A full set of waiting times covered the range from 6×10^2 s (time resolution of field observations) to $\sim 1.3\times10^3$ s (waiting time for the most rare events). The waiting times for each arbitrary chosen value A were plotted in doubly logarithmic coordinates against A (Figure 10). In general, the trend is obvious: the higher amplitude, the longer waiting time. However, in the range of stronger accelerations $1\times10^{-5} < A < 2.5\times10^{-5}$ the distribution function $\tau(>A)$ fits the straight line that represents the power law:

$$\tau(> A) \propto A^{\gamma} \qquad (5)$$

This dependence means the temporal invariance of the drift for, at least, a decade of time (a whole period of observations covered two orders of magnitude on the time scale). Consequently, the power law behaviour (which is indicative of critical dynamics) requires the presence of sufficiently strong forcing.

The deviation from the power law dependence for the strongest accelerations, which require the longest time of expectation, is due to the insufficient statistics for the most dramatic events. As regards the effect of lower correlation of smaller events, one should take into account that the terms “small” and “large” as referred to the events in the dynamic process are relative definitions. The events should be compared not between each other but in relation to the size of the entire system. This problem was studied analytically by Christensen and Olami (1992) and discussed again by Sánchez et al. (2002). Both groups of authors concluded that owing to the natural attenuation in real geophysical systems, a small event occurring in one part of the system cannot feel the presence of another one. At the same time, if one studies the process in a sufficiently limited subsystem, the correlations between former "small" events would manifest itself.

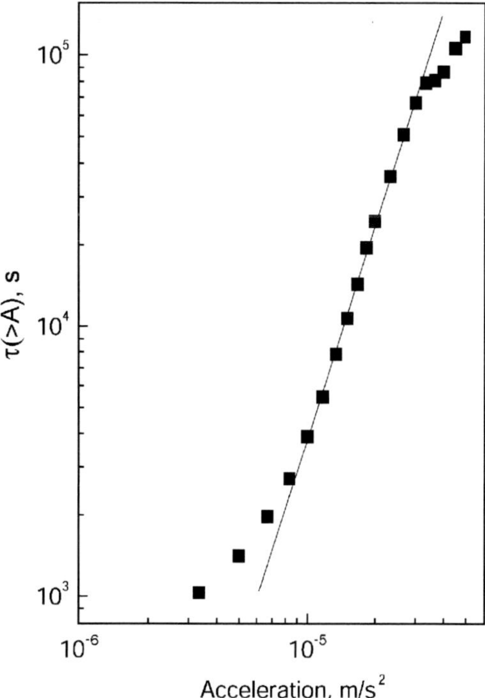

Figure 10. Waiting times of accelerations with amplitudes higher than given by the corresponding abscissa. An original sequence of accelerations is depicted in Figure 8. The straight line fits the power law (5).

This conclusion was supported by a similar analysis of waiting time distribution performed for the accelerations averaged over much longer time intervals. Figure 11 shows the accelerations of ice-field measured in the NP 31 during one year. In this case, the accelerations of the ice field were calculated on the base of the velocity changes measured one time a day (ΔV = 1 day). Correspondingly, the amplitudes of accelerations were significantly smaller then that averaged over 10 min intervals given in Figure 7. Nevertheless, the power law dependence (5) remains valid for the time intervals between "smaller" events (Figure 12). One should conclude that the temporal invariance of the ice field motion does not depend on the scale of observation.

At the same time, the values of the exponent γ determined from the slope of log-linear portions in Figure 10 and Figure 12 are significantly different. This means that the scaling properties depend on the scale of events. The problem of the universality of time-scaling parameters in geophysics was considered in detail by Davidsen et al. (Davidsen and Goltz, 2002; Davidsen et al., 2006) who have shown that the common function for scale-invariant waiting time distributions of events of the same nature could be found taking into account the spatial and/or energy scale of events. Bak et al. (2002) pointed out that the universal power law for waiting distribution can be constructed only when taking into account the spatial and energy importance of the events under consideration.

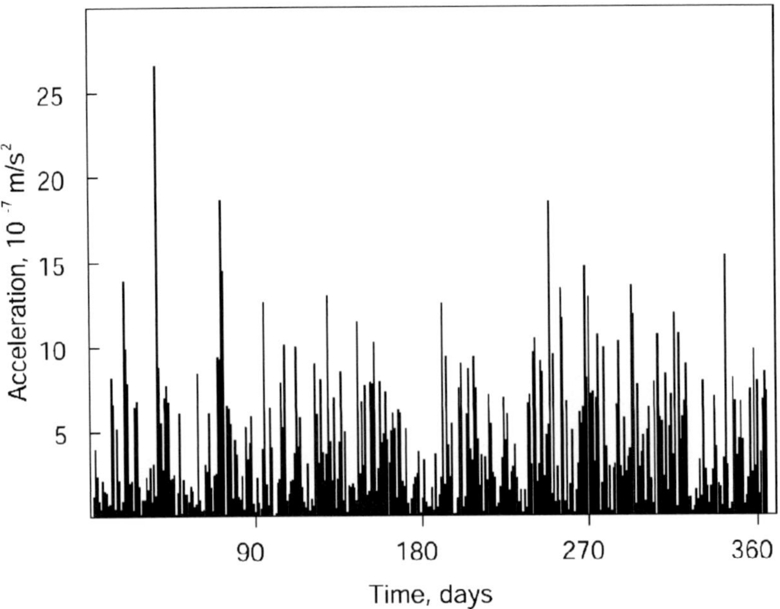

Figure 11. Amplitudes of accelerations measured in the period of time from 25 October 1988 to 24 October 1989. The accelerations are referred to daily changes of the ice field velocity. From the database of NP 31.

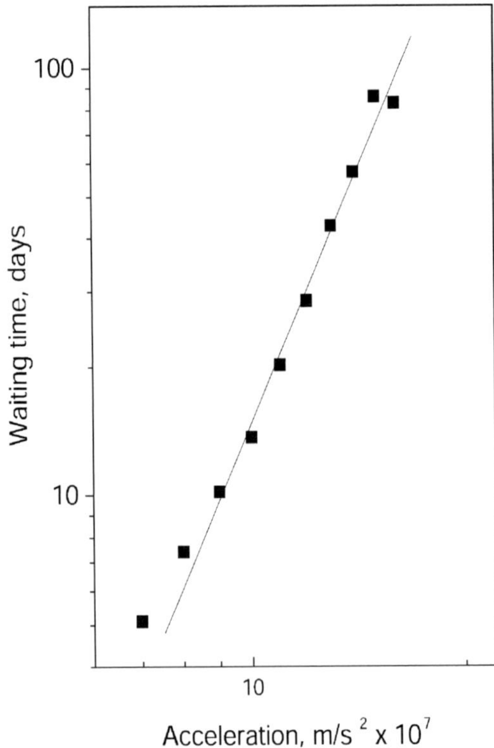

Figure 12. Distribution of waiting time of accelerations with respect to their amplitudes. A sequence of accelerations is depicted in Figure 11.

To summarize this Section, the power law time sequence of events in the statistical system evidences the presence of a long term memory of its evolution. In our case, each new event of acceleration is affected not only by the most recent events but by the long sequence of previous force interactions. Thus, the current and future behavior of drifting sea-ice is determined by its pre-history. The property of memory implies a certain degree of conservation in the system, since in a system the events are independent on each other. The role of the conservation is closely related with the energy characteristics of the sea- ice dynamics, which we shall consider in the next Section.

4.2. Energy Exchange

The dynamics of the sea-ice drift is determined, on the one hand, by the interaction of mobil components of the ASIC with the atmosphere (wind, pressure) and ocean (currents, tides), and, on the other hand, by the internal interaction between ice sheets. All these interactions cause the energy exchange, which, in turn, produces the dynamic integrity of the ASIC over the basin-wide area. The scaling properties of the energy exchange could also be assessed from the data on the motion of individual ice-fields, though this approach does not allow one to distinguish between changes in kinetic energy due to either direct external forcing or mechanical interactions between neighboring components.

The energy release, E, is directly proportional to the change in the speed squared of the field, $E \propto \Delta V^2$. Therefore, the distribution function of the energy release is equivalent to the speed distribution function $N(>\Delta V^2)$ (here N is the number the events of speed change squared exceeding the given value ΔV^2). A series of velocity changes in 10 min intervals was measured during the period of time from 1 January to 28 February 2004. This time interval was selected for statistical analysis since it covered both the period of low atmospheric forcing and the period of drift under conditions of the deep atmospheric depression with corresponding significant changes in the temperature and wind velocity (Figure 13). In addition, on 10 February 2004, a multiple fragmentation of the sea-ice in the region of the NP 32 drift was detected both in the research station itself (records in the log-book), and, a posteriori, in the NOAA satellite images (Figure 14).

The functions $N(>\Delta V^2)$ were found separately for three periods of observations: (i) from 1 January to 31 January ("remote" period); (ii) 1 February to 10 February ("pre-catastrophic/catastrophic" period); (iii) 11 February to 28 February ("posterior" period). Figure 15 shows the obtained distributions plotted in log-log coordinates. In all three plots one can see portions approximated with straight lines that represent the power law dependences:

$$N(>\Delta V^2) \propto (\Delta V)^{-b} \tag{6}$$

Hence, for the energy release we have a scaling relation:

$$N(>E) \propto E^{-b} \tag{7}$$

The b-values calculated from the slopes of straight lines were found to be equal to 2.12 ± 0.08 in the "remote" and "posterior" periods, with b = 1.51 ± 0.06 in the "pre-catastrophic" period. Its variation during different periods of drift means that the b-value is not a universal parameter of the scale-invariant energy exchange. What does signify its variation? From the formalistic viewpoint, the slope of straight lines characterizes the relative contribution of "large" and "small" events. The challenge is to bind the revealed alternation of the prevailing scale of events with general conditions of the process.

This problem has a long pre-history in geophysics. There were reported many evidences that the decrease of the b-value in series of energy-release events related to seismic activity could serve as a pre-fracture signature (Kapiris et al., 2004 and references therein). In our case, a reversible drop of the b-value from 2.12 to 1.51 occurred also prior to the large scale sea-ice fragmentation. Therefore, in order to elucidate the physical sense of the b-value, and to interpret its variation, we shall consider the premises of its introduction to the geophysics and discuss the thermodynamic significance of this parameter.

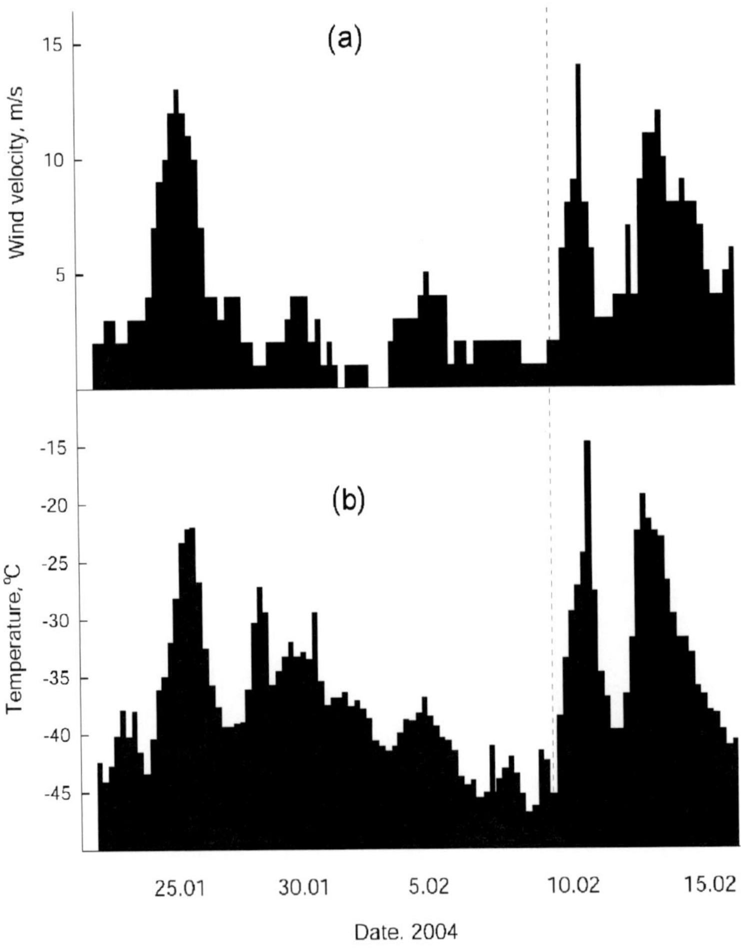

Figure 13. Histograms of the wind velocity and the air temperature in the period of time covering the event of 10 February 2004 (Figure 14). Dashed line indicates the day of 10 February.

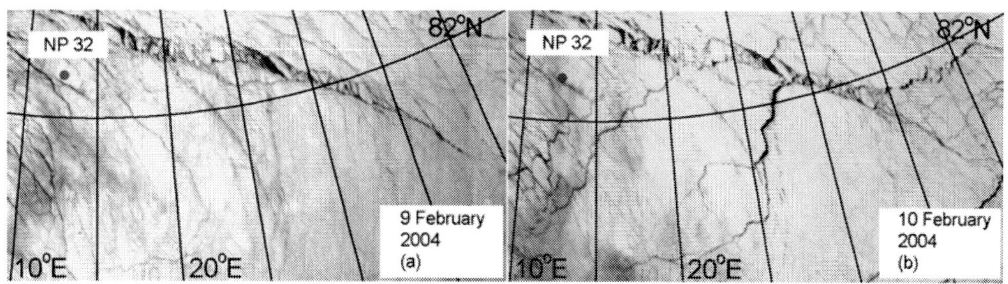

Figure 14. 400 × 200 km fragments of NOAA satellite images of the region of NP 32 drift obtained on 9 February (a) and 10 February (b) 2004. A newly formed lead is seen in (b).

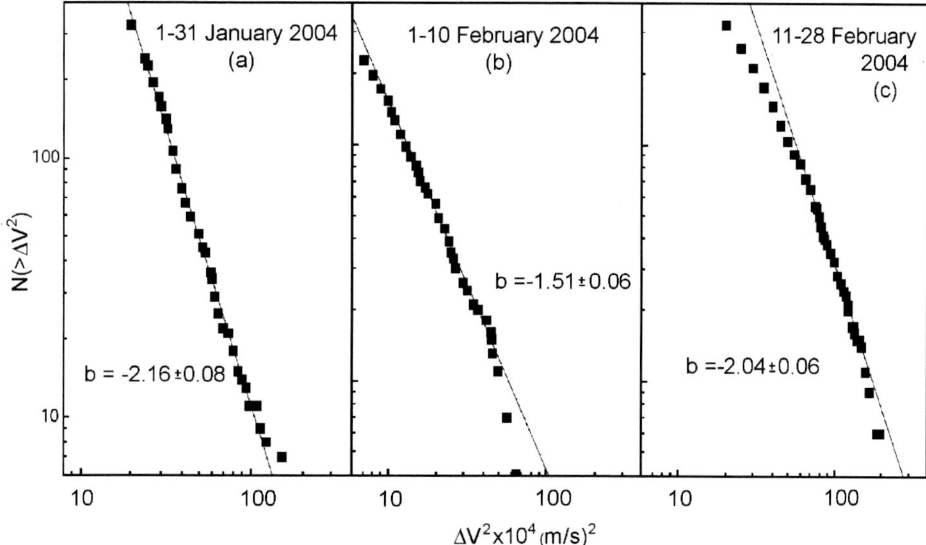

Figure 15. Distributions N>ΔV^2 versus ΔV^2 in three periods of field observations.

The parameter b appeared firstly as a log-linear constant in the formulation of the statistical distribution of magnitudes (M) of earthquakes given by Gutenberg and Richter (1956):

$$\log_{10} N(>M) \propto - bM \quad (8)$$

where N is the number of earthquakes whose magnitude is higher than M. The magnitude M is proportional to the logarithm of the released energy with a factor d that varies between 1 and 1.5 (Kanamori, 1978; Ekström and Dziewonski, 1988). Therefore one could rewrite the relation (8) in a form of the following power law (Kanamori, 1978):

$$N(>E) \propto E^{-b/d} \quad (9)$$

where the N(>E) is the number of earthquakes with the energy greater than E. As a rule, the presence of hardly determined parameter d in the power index is ignored in the literature, and the scaling law (9) is usually referred in the form (7). The estimates for the b-value given by different researchers on the base of both natural (earthquakes) and laboratory (rock fracture) observations fall on the range 0.6 to 1.5. A particular value of this parameter depends on physical and mechanical properties of an object, while the variation characterizes indirectly the stage of the pre-fracture process (Enescu and Ito, 2001).

The variation of the b-value during a permanent dynamic process remains an unresolved problem up to now. There were put forward a variety of mechanical models that would explain the decrease of the b-value before large-scale fracture events. A brief (but informative) review of the significance of various parameters that could, as suggested, control the b-value was given by Amitrano (2006). The main trend is the following (Mori and Abercombie, 1997; Sue et al., 2002): the lower heterogeneity of fracturing material, the lower b-value. However, up to now no structural models were developed. This is an indirect consequence of the principal postulate of the statistical conjecture that the scale-invariant processes require a structure-free interpretation.

Really, on the one hand, all changes in the statistical parameters are caused by the interplay of physical conditions and interactions. This is particularly true for the case of open non-equilibrium systems which is under consideration in this chapter. On the other hand, the space-time fractal properties of the system does not depend on particular structural links or timing of events. Therefore, the explanation of the behavior of scaling parameters, such as the b-value, one should search in general physical principles.

This approach was realized by Christensen and Olami (1992; 1992a) and Olami et al. (1992) who, in order to gain insight into the mechanism that governs the b-value variation, developed a two-dimensional spring-block model where the energy release distribution in fracture events was related with the conservative properties of the non-equilibrium system driven by the outer force; a level of conservation was defined through a ratio between the elastic constants. Their main finding was the following: the higher the level of conservation, the smaller the b-value. In a fully non-conservative system the scaling is impossible due to the lack of interaction between its components. It should be stressed that the particular mechanism that controls the level of system's conservation is of low significance: first and foremost is the character of the system's interaction with the energy flow that passes through.

In this light, the decrease of the b-value observed in the period prior to the large-scale sea-ice fragmentation evidences the increase of the conservation of the system of drifting sea-ice. The physical cause of the change of conservation could be a substantial decrease of the air temperature (down to –35 to -45°C) that had taken place in the "critical" period (Figure 13b). The temperature drop gained freezing with enhancing the connectedness of the sea-ice cover. The range of strain correlation in more consolidated substance is higher due to higher interconnection between components. The divergent scale of correlation signalizes the criticality of the system's state (Salminen et al., 2002).

At the same time, the increased energy conservation in the consolidated sea-ice leads to the decrease of the total amount of the energy-release events typical for "quite" periods of the sea-ice dynamics. A lower slope of the $\log N(>\Delta V^2)$ versus $\log(\Delta V^2)$ dependence in the time interval from 1 February to 10 February 2004 (Figure 14) indicates the relative decrease of the contribution of smaller events to the total amount of detected velocity changes. This is a reasonable effect of the suppression of "fine" motion in the frozen sea-ice cover.

5. Self-Organized Criticality

In contrast to truly random processes in the thermodynamically equilibrium systems where the fluctuations are small, and, correspondingly, the correlations lengths and times are also small and decay exponentially, in the non-equilibrium systems with a great number of metastable states, the size of fluctuations can spontaneously increase without a fine tuning of the external parameters. The correlation radius in the latter case is a power law function typical for fractal structures.

Fractals are not associated with any characteristic length, that is, they are critical features in the sense defined in Section 4. The criticality emerges due to the self-organized dynamics that determines the behavior of the entire system. Although the attractor of the dynamics (pattern of fragmented cover) is marginally stable, the dynamic irregularities (fluctuations or avalanches of various size) cause the redistribution of the energy passing through the system, so that the irreversible global failure never occurs. This phenomenon was called the "self-organized criticality" (SOC) (Bak, Tang and Wiesenfeld; 1993).

All attributes of the SOC, such as the spatial, temporal and energy invariance that manifests itself by relevant power-law functions, are inherent in the ASIC. The externally-driven dynamics (atmosphere and ocean forcing) is certainly non-equilibrium. However, the mechanism, that controls the long-range interaction of sea-ice cover fragments and the "memory" about preceding events remains shaded since the driving force of the self-organization cannot be determined in the framework of the statistical physics (excluding general thermodynamical principles (Arneodo, 1995)). A formalistic description of events restricts the significance of the analysis of the behavior of sea ice exhibiting the features of the self-organization. Therefore, in order to establish a casual-effect relation between a multiplicity of individual mechanical events in the ice pack and their common fractal properties, one should search the physical processes, which could result in the correlated dynamics of the ASIC.

The problem of "physical carrier" of the self-similarity is the main challenge when applying the SOC concept to any real non-equilibrium process. The fractal properties of an object do not depend neither on its structure, nor on the nature of the energy exchange between components but to maintain the scale invariance, should the whole system be interconnected dynamically. In the case of ASIC, the dynamic connectedness of the collection of sea-ice sheets is realized through the wave and oscillation processes, which transfer and redistribute the energy income in the sea-ice cover.

6. Oscillation Processes

The ASIC is the complicated oscillation system. The pack oscillations and waves of various nature and origin arise and extend in the sea ice permanently and simultaneously in all parts. The ASIC is subjected, firstly, to the action of the sea gravity waves propagating from the areas of open water, and tidal waves. Secondly, the spectrum of outside coming waves is supplemented with the variety of elastic waves excited in mobile, stressed, breaking-and-freezing sea-ice formations. Each relative displacement between ice pieces is accompanied by bursts of oscillations of various kinds whose propagation length is limited by the attenuation effects and boundary conditions. The drifting ice fields, when interacting, transfer to each other pulses (or trains of pulses in the case of, for example, stick-slip passages) of positive/negative accelerations. This process is easily available for instrumental measurements carrying out on the ice pack. In contact interactions the amplitudes of the horizontal displacements are much higher than the amplitudes of the vertical components (Martin and Drucer, 1991). Therefore, using the horizontal accelerometers established in the sea-ice field one can monitor this process in detail (Martin and Becker, 1988; Smirnov, 1996; Smirnov, 2001). The difference between vertical and horizontal accelerations is clearly seen in Figure 16 where the stick-sleep process was recorded simultaneously by three differently oriented accelerometers established on the sea-ice sheet slowly moving along the crack. The larger amplitudes of the X- and Y-oriented oscillations indicate their direct relation with the horizontal motion of sea-ice fields.

An example of isolated pulses of acceleration caused by the impact interaction between ice-fields is depicted in Figure 17. The interactions of this kind could be a part of the more complicated oscillation process (Figure 18), whose spectrum could be composed of periodic motions due to ripple and those due to shear and impact interactions accompanied with hummocking and cracking.

The oscillation processes in the sea ice are highly affected by the rheological properties of this medium. Under the prolonged (non-impact) action of the compressive/tensile force, a feedback between the energy input and the deformation properties of the oscillating domain composed of adjacent sheets takes place; this effect causes the parametric redistributing of the vibrational energy.

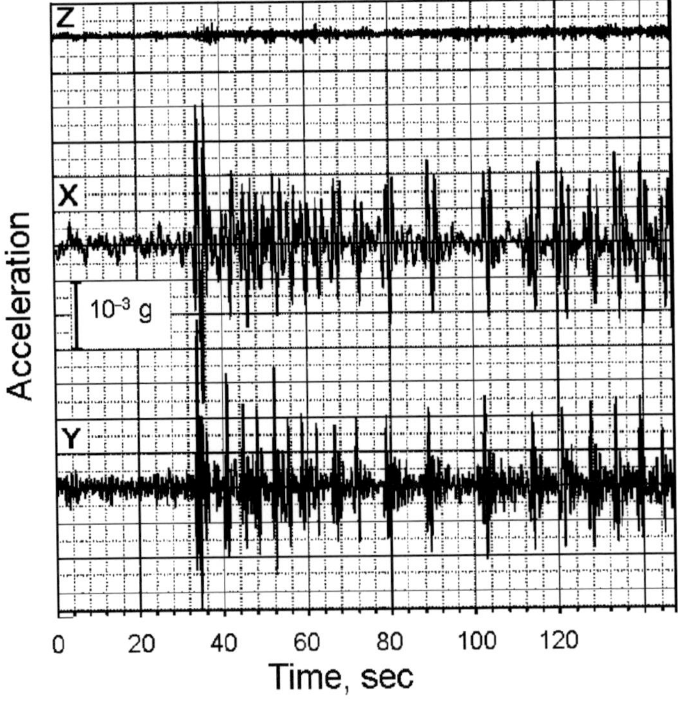

Figure 16. Shearing along the collapsed crack (slip-stick). Three accelerometers were oriented along different horizonatal (X and Y) and vertical (X) axes. From the database of NP 35 (2007).

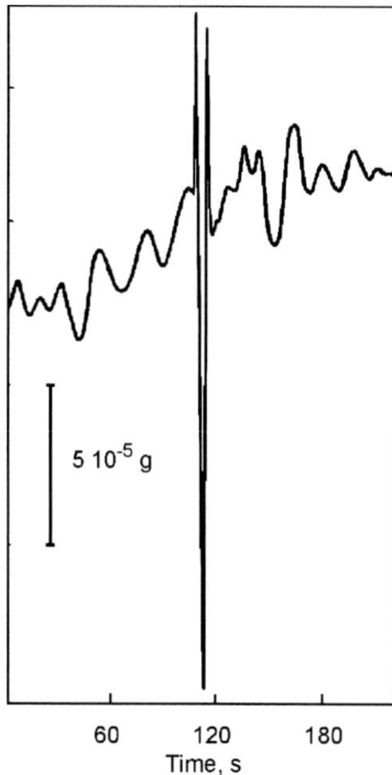

Figure 17. An isolated pulse of the impact interaction between sea-ice fields. From the database of NP 22 (1975).

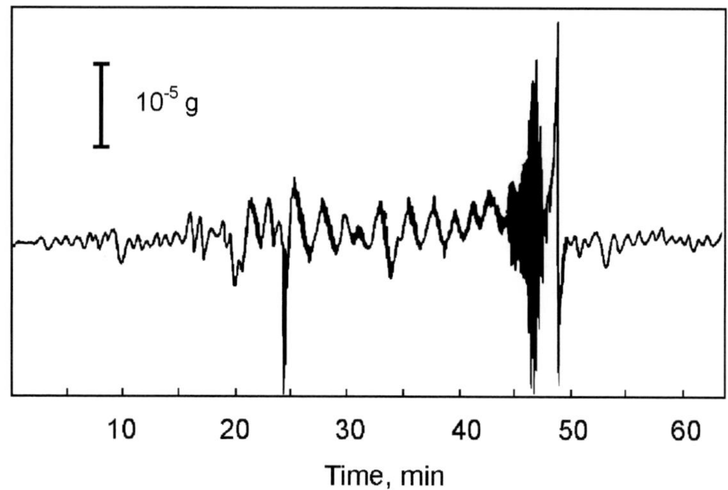

Figure 18. Oscillations of moving ice sheet ("sea-ice island") excited by its interaction with consolidated ice pack. From the database of NP 23 (1977).

As a result, the period and amplitude of oscillations do not depend on the existence of periodicity in the outer forcing, and are determined by the intrinsic properties of the vibrational system. The parametric interaction leads to so-called self-oscillations. In contrast to the forced oscillations under the action of periodic force, the energy income to the self-oscillation system from a non-periodic external source is dosed by the system itself. The shape and the periodicity of self-oscillations depend on the speed of the crack wall displacement. At relatively low speed, the regular non-linear pulses prevail, while at higher speed the oscillations are quasi-harmonic (Figure 19). The self-oscillation process continues until the energy income from the outer source is not depleted. Thus, the energy of compression/tension transforms permanently to the correlated motion of ice sheets.

In most cases, each ice sheet is involved simultaneously in various interactions with adjacent formations, such as other sheets, consolidated ice pack and shore structures (including marine engineering facilities). The compression and shearings are related with different modes of deformation and fracture of contact zones in the sea-ice. These interactions are reflected in the spectrum of oscillations by complicated trains of pulses and multi-frequency excitations (Figure 20).

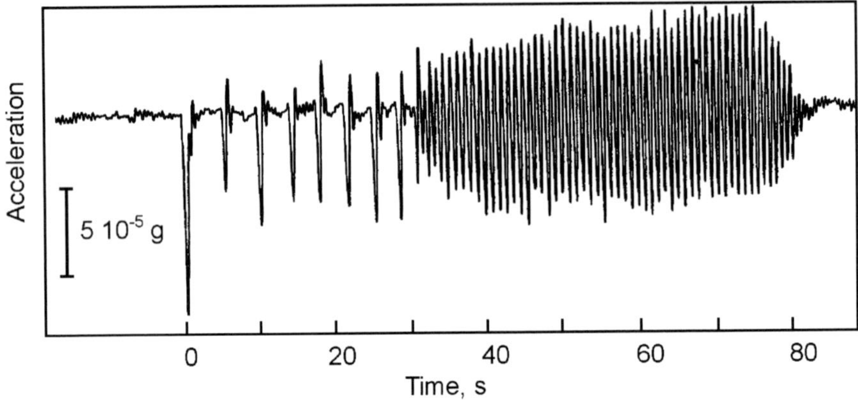

Figure 19. Self-oscillations due to shearing along the collapsed crack. The periodic "jumps" observed during first 30 sec transform to sliding of crack walls. From the database of NP 20 (1970).

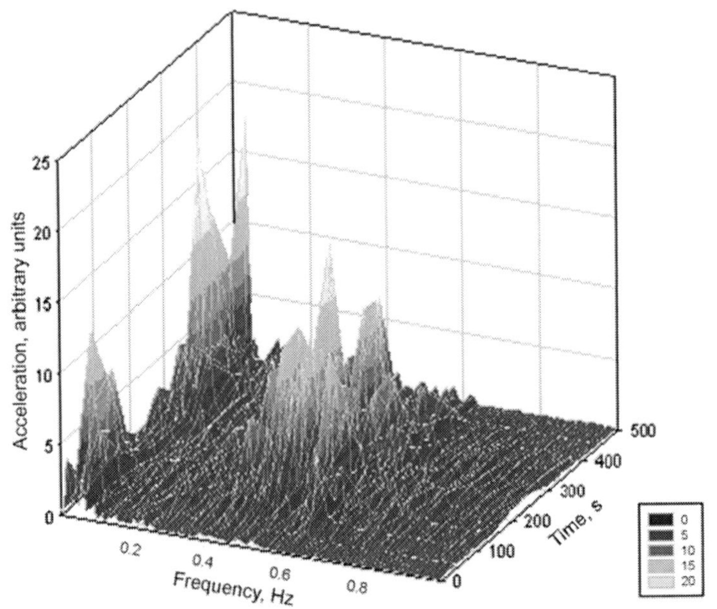

Figure 20. Spectrum of oscillations induced by a complex horizontal displacements of the ice-field. From the database of NP 33 (2005).

No relations with meteo-factors were detected in this or other similar records; consequently, the observed self-oscillations emerge as a self-organizing process under the permanent (non-periodic) external force. Synchronous observations carried out using sensors established on adjacent ice fields showed that the correlation radius of such phenomena could exceed the dimensions of individual pack formations with involving into the correlated motion important areas (Smirnov, 2001).

The expanded energy exchange through oscillations and waves provides the feedback mechanism needed for maintaining the critical state. The horizontal ice shearings result in self-oscillations of different spatial and temporal scales, which provide the local stress release; the whole system returns to the equilibrium/critical state until another large-scale event occur. Thus, the spectrum of cyclic shearings of ice reflects a variety of self-similar processes in the permanently critical system. The parametric interactions of oscillations provide conditions for the energy re-distribution over vast areas of the sea-ice (with open water inclusions), what is the requirement for self-organizing of the system.

CONCLUSION

The satellite images of the Arctic sea-ice cover manifest the fractal geometry of the crack-lead-ridge pattern. The analysis of the ice sheet displacements that take place during the pack drift evidences the time invariance of the drift dynamics. The sea-ice cover evolves with keeping valid the scale-independent (power law) relations between the spatial, temporal, and energy parameters of drift-related processes. These scaling properties allow one to consider the ASIC as the self-organizing, space-time-energy fractal domain.

The lack of any characteristic lengths and times in the drift dynamics indicates the permanent criticality of the ASIC. The extended occurrence of fractal structures in nature evidences the great role of critical states in the spontaneous evolution of the complex open systems including the geophysical systems. This self-organized criticality is the particular, marginal (i. e. close to the instability threshold) state to which the system goes through an infinite sequence of events of different dimension and different duration. The long periods of quasi stationary, scale-invariant evolution alternate with the short-term "catastrophic" burst of events covering the whole system or its significant parts. (Chelidze et al, 2006; Chmel et al., 2007).

Correspondingly, the scaling properties of the sea-ice drift dynamics are sensitive to large-scale perturbations. The fractal dimension of the collection of fragments limited by leads, cracks and ridges decreases in response to the fracture events in agreement with well-established in geophysics trend of decreasing the dimensionality of the fracturing system prior to fault nucleation. (A term "fault" in this context is applicable to giant branching leads like those seen in Figure 1 (Weiss. 2003)). The variation of the b-value in the period preceeding to large fragmentation signalizes the preponderance of more important events. These findings seem to be of practical significance as they could be potentially useful for forecasting the events of geophysical scale issuing from the variation of the structure-independent scaling parameters in addition to searching the trends in real physical processes (Chmel and Smirnov, 2007).

The mechanism of the long-range correlations in the sea-ice cover is related with the self-organized dynamics of oscillations and waves inherent in the ice pack.

References

Aksenov, Ye. In *Local ice cover deformation and mesoscale ice dynamics — Reports*, Editors Ryska, K. and Tuhkuri, J., 1999; Part 1. Picaset Oy Espoo, Helsinki, p. 100-147.

Amitrano, D. *Intern. J. Fract.* 2006, 139, 369-381.

Arneodo, A.; Bacry, E.; Muzy, J.F. *Physica A* 1995, 213, 232-275.

Bak, P.; Christensen, K.; Danon, K.; Scanlon, T. *Phys. Rev. Lett.* 2002, 88, 178501 (1-4).

Bak, P.; Tang, C. *J. Geophys. Res.* 1989, 94, 15635-15637.

Bak, P.; Tang C.; Wiesenfeld K. *Phys. Rev. Lett.* 1987, 59, 381-384.

Bouchaud, E. *J. Phys.: Cond. Matter* 1997, 9, 4319-4344.

Caldarelli, G.; Di Tolla, F.D.; Petri, A., *Phys. Rev. Lett.* 1996, 77, 2503-2506.

Chelidze, T.; Kolesnikov, Yu.; Matcharashvili, T. *Geoph. J. Intern.* 2006, 164, 125-136.

Chmel, A.; Kuksenko, V.S.; Smirnov, V.N.; Tomilin, N.G. Nonlin. *Processes Geophys.* 2007, 103-108.

Chmel, A.; Smirnov, V.N. *Physica A*, 2007, 375, 288-296.

Chmel, A.; Smirnov, V.N.; Astakhov, A.P. *JSTAT,* 2005, P02002 (1-11).

Chmel, A.; Smirnov, V.N.; Panov, L.V. *Ocean Sci.* 2007, 3, 291-298.

Christensen, K.; Olami, Z. *J. Geophys. Res.* 1992, 97, 8729-8735.

Christensen, K.; Olami, Z. *J. Phys.Rev. A* 1992a, 46, 1829-1838.

Enescu B., Ito K. *Tectonophys.* 2001, 338, 297-314.

Grassberger, P.; Procaccia, I. *Phys. Rev. Lett.* 1983, 50, 346-349.

Kapiris, P.G.; Balasis, G.T.; Kopabas, J.A.; Antonopoulos, G.N.; Peratzakis, A.S.; Eftaxias, K.A. *Nonlin. Proc. Geophys.* 2004, 11, 137-151.

Korsnes, B. *Statistical description and estimation of ocean drift ice environments*, 1991; Norges Tekniske Høgskole, Trondheim, pp. 1-19.

Korsnes, R.; Souza, S.R.; Donangelo, R.; Hansen, A.; Paczuski, M.; Sneppen, K. *Physica A* 2004, 331, 291-296.

Lewis, J.K.; Tucker, W.B.; Stein, P.J. *J. Geophys. Res.* 1994, C99, 16361-16371.

Mandelbrot, B.B. *Intern. J. Fracture* 2006, 138, 13-17.

Mandelbrot, B.B.; Passoja, D.F.; Paullay, A.J. *Nature* 1984, 308, 721-724.

Marsan, D.; Stern, H.; Lindsay, R.; Weiss, J. *Phys. Rev. Lett.* 2004, 93, 178501 (1-4).

Martin, S.; Becker, R. *J. Geophys. Res.* 1988, 93, 1303-1315.

Martin, S.; Drucer, R. *J. Geophys. Res.* 1991, 96, 10567-10580.

Maslov, S.; Paczuski, M.; Bak. P. *Phys. Rev. Lett.* 1994, 73, 2162-2165.

Mori, J.; Abercombie, R.E. *J. Geophys. Res.* 1997, B102, 15081-15-90.

Olami, Z.; Feder H.J.S.; Christensen, K. *Phys. Rev. Lett.* 1992 68, 1244.

Salminen, L.I.; Tolvanen, A.I.; Alava, M.J. *Phys. Rev. Lett.* 2002, 89, 185503 (1-4).

Sánchez, R.; Newman, D.E.; Carreras, B.A.; *Phys. Rev. Lett.* (2002) 88 68302 (1-4).

Smirnov, V.N. *Dynamic Processes in sea ice, 1996*; Gidrometeoizdat, St. Petersburg, Russia, (In Russian).

Smirnov, V.N. *Proc. 16-th Intern. Conf. on Port and Ocean Eng. under Arctic Conditions*, POAC, Ottawa, 2001, 421-429.

Sue, C.; Grasso, J.R.; Lahaie, F., Amitrano, D. *Geophys. Res. Lett.* 2002, 29, 65 (1-4).

Tucker, W.B.; Perovich, D.K. *Cold Region Sci. Technol.* 1992, 20, 119-139.

Weiss, J. *Surv. Geophys*. 2003, 24, 185-227.

Weiss, J.; Marsan, D. C.R. *Physique* 2004, 5, 735-751.

In: The Pacific and Arctic Oceans
Editor: Kallen B. Tewles

ISBN: 978-1-60692-010-7

Chapter 5

SEA ICE DRIFT IN THE ARCTIC OCEAN: SEASONAL VARIABILITY AND LONG-TERM CHANGES

V. K. Pavlov and O. A. Pavlova
Norwegian Polar Institute, Polar Environmental Centre,
Tromsoe, Norway

ABSTRACT

Variability in the drift of sea ice in the Arctic Ocean is an important parameter that can be used to characterise the thermodynamic processes in the Arctic. Knowledge of the features of sea ice drift in the Arctic Ocean is necessary for climate research, for an improved understanding of polar ecology and as an aid to human activity in the Arctic Ocean. Monthly mean sea ice drift velocities, computed from Advanced Very High Resolution Radiometer (AVHRR), Scanning Multichannel Microwave Radiometer (SMMR), Special Sensor Microwave/Imager (SSM/I), and International Arctic Buoy Programme (IABP) buoy data, are used to investigate the spatial and temporal variability of ice motion in the Arctic Ocean and Nordic seas during the 27 years from 1979 to 2005. Sea ice drift in the Arctic Ocean is characterized by strong seasonal and inter-annual variability. Sea ice drift velocities mirror seasonal changes of the wind in the Arctic, reaching a maximum in December, with a minimum in June. In the central part of the Arctic Ocean and in the area near the Canadian shore the amplitude of this variation is not more than 2 cm s^{-1}. The maximum amplitudes are found in the Fram Strait (9-10 cm s^{-1}), Beaufort Gyre (6-7 cm s^{-1}) and the northern part of Barents Sea (5-6 cm s^{-1}). Low frequency variations of sea ice drift velocities, with periods of 2.0-2.5 yrs and 5.0-6.0 yrs, are related to reorganization of the atmospheric circulation over the Arctic. There is evidence that the average sea ice velocity for the whole of the Arctic Ocean is increasing, with a positive trend for the period 1979-2005. Trends of the monthly mean ice drift velocities are positive almost everywhere in the Arctic Ocean. In the Baffin Bay, Fram Strait and Barents Sea regions, sea ice velocities have increased dramatically, by up to 0.15-0.20 cm s^{-1} per year. We suggest that the increase of the sea ice drift velocities in the Arctic Ocean is mostly related to the decrease in both sea ice concentration and ice thickness. The character of the inter-annual variability of ice exchange between the marginal Arctic seas and Arctic Basin and in the Fram Strait in general is similar to the variability of sea ice drift velocities averaged for the whole Arctic Ocean. We have found an increase (50-70%) in the ice export from Siberian seas to the Arctic Basin.

INTRODUCTION

Sea ice is a dominant feature of the Arctic Ocean and its marginal seas. Sea ice extent has usually ranged from a maximum in March of about 16 million km^2 in March to a minimum of 7 million km^2 at the end of the summer melt season, in September. Dramatic changes in sea ice conditions of the Arctic Ocean have been seen during the past three decades (satellite era) and these changes have been related to warming and reorganization of the atmospheric circulation. Sea ice has decreased during the period 1979-2007. The linear trends in Arctic sea ice extent are negative for every month. Analyses of passive microwave satellite data from 1979 to the present indicate significant declines in maximum and minimum ice extents and areas, and the minimum values decrease about 5 times faster than the maximum values. The most rapid decrease is for September, when the trends in summer minimum extent and area are respectively about -8.6 % and -9.6% per decade. The corresponding values for winter maximum extent and area are -1.9% and -2.5% (Comiso, 2006; Serreze, Holland and Stroeve, 2007, Gerland et al., 2007). Arctic sea ice has been subject to changes in the past decade, with minimal ice extent during summers of 2002, 2005 and 2007, as measured from satellite microwave data (Serreze et al., 2003; Francis et al., 2005; Stroeve et al., 2005; Comiso et al., 2008). Every year since 2001 has pronounced September minima, the most extreme of which was in September 2007 when the extent and area of the sea ice cover reached at 4.1×10^6 km^2 and 3.6×10^6 km^2 respectively. These are 24% and 27% lower than the previous record low in September 2005, and 37% and 38% less than the climatological averages (Stroeve et al., 2005; Comiso et al., 2008). Significant changes in the motion of ice in the Arctic Ocean have also been detected (Rigor, Wallace and Colony, 2002; Pfirman et al., 2004; Martin and Gerdes, 2007; and others).

Sea ice in the Arctic Ocean is constantly drifting and being deformed, redistributed and broken into fragments. Externally driven forcing includes winds, ocean currents, tides and sea level gradients. Sea ice drift (SID) changes sea ice concentration (SIC), creating gaps in the ice cover, thus providing the preconditions for new ice formation, which in turn results in increasing the sea water salinity. SID redistributes solid freshwater and latent heat energy, making a significant contribution to the freshwater and heat budgets of the Arctic Ocean and adjacent areas of the Atlantic and Pacific oceans, possibly affecting the global thermohaline circulation (Mauritzen and Hakkinen, 1997; Lohmann and Gerdes, 1998). Thus SID plays an important role in the formation and maintenance of the Arctic climatic system.

The study of SID is one of the most important lines of inquiry in polar ecology. Drifting ice carries with it pollutants from the region of its formation, which results in a partial cleansing of this region and in pollution of the region where the ice melts (Pavlov and Stanovoy, 2001; Pavlov, 2006). The investigation of SID is of practical importance for the development of a theory of the mechanisms of pollutant and sediment transport (Weeks, 1994; Pfirman et al., 1995; Pfirman, Kogler and Rigor, 1997a; Chernyak, Rice and McConnell, 1996; Pavlov and Stanovoy, 2001; Korsnes, Pavlova and Godtliebsen, 2002; Pavlov, Pavlova and Korsnes, 2004; Pavlov, 2006; and others).

Ice floes are also used as platforms for making various types of observations in the open Arctic Ocean. Examples of such activity are Russian drifting stations, American ice floe stations, drifting buoys and the recent international TARA expedition, carried out in the framework of the DAMOCLES Project (http://www.taraexpeditions.org).

Thus knowledge of the specific features of SID in the Arctic Ocean is necessary for climate research, for an improved understanding of polar ecology and will also aid human activity in the Arctic Ocean.

In this study we investigate the changes of ice motion in terms of SID velocities and show how atmospheric processes, SIC and ice thickness affect the SID variability.

A Brief History of Sea Ice Drift Studies

The study of sea ice in the Arctic Ocean and Nordic Seas has a long history. Before the 19th century the study of the Arctic was of an episodic character. Expeditions were mostly dispatched to search for new lands, such as the voyages by the Vikings in the 9th-13th centuries to the shores of Iceland, Greenland, America and White Sea; and voyages in the 11th-16th centuries by Russian Pomors to Spitsbergen-Grumant and Novaya Zemlya.

From the beginning of exploration and travel within Arctic seas, the main problems encountered were related to poor knowledge of the ice and ocean's conditions. The use of Arctic sea routes, and exploration and development of the region's natural resources were impossible without understanding the typical spatial and temporal (seasonal and inter-annual) variations in the atmosphere, ice cover. It was necessary to organize systematic observations of weather, ice and water over the entire Arctic.

In 1879 US naval officer George W. De Long headed an expedition to the North Pole on board the ship Jeannette (De Long, 1883). Jeannette departed San Francisco on July 1879 and through the Bering Strait reached the Chukchi Sea. In September the same year she was caught fast in the ice pack near Wrangel Island. For the next 21 months Jeannette drifted to the northwest, ever-closer to DeLong's goal, the North Pole itself. In June 1881 the pressure of the ice finally began to crush Jeannette and she went down north of New Siberian Island. Artefacts from ship were later found in the ice off southwest Greenland in 1884. Based on these finds, the Norwegian meteorology professor, Henrik Mohn, suggested the existence of a current from the shores of East Siberia to the passage between Spitsbergen and Greenland, in other words, a constant east-west current across the Arctic Ocean, later named Transpolar Drift (TPD) Stream.

To prove this hypothesis Fridtjof Nansen, a famous Norwegian traveller, scientist and public man planned and executed the Fram expedition (Figure 1). He also recognized in Mohn's theory the only practical way of reaching the North Pole and the possibility of simultaneously carrying out extensive studies of the near-Pole area of the Arctic Ocean.

Nansen conceived a plan to build a special ship which could withstand the force of the ice. According Nansen's plan such ship had to be intentionally frozen into the ice as far to the north as possible of where the Jeannette had gone down, and left to drift westward with the TPD, hopefully across the North Pole. Nansen called his ship the Fram ("Forward" in Norwegian), and on 24 June, 1893 the Norwegian Polar Expedition started. The Fram drifted no further than 86° N and the most northerly position (85°56'N, 66°31'E) was reached on 15 November 1896.

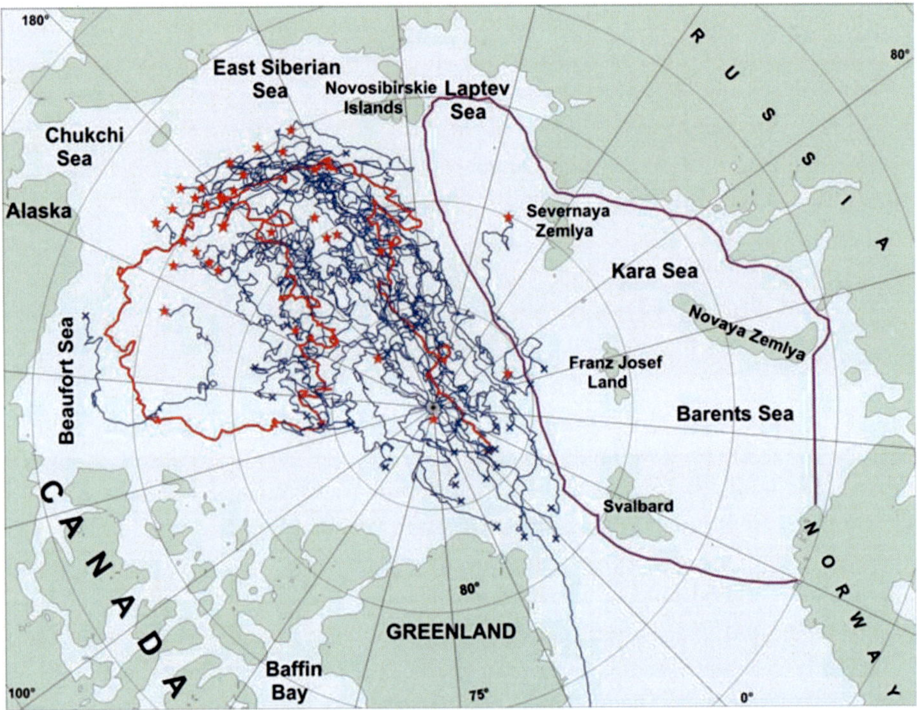

Figure 1. Map of the Arctic Ocean with the drift tracks of the North Pole (NP) Stations for the period 1937-2008 (blue lines), and the course and drift of the Fram from 1893-96 (violet line). Red line-the drift of NP-22.

At that point the drift direction changed to the south-west. Nansen's expedition was completed on 23 August 1896. The expedition lasted for three years (1893-1896), and extensive meteorological and oceanographic observations in the Arctic Basin were carried out (Nansen, 1903).

The drift of the Fram confirmed the validity of Mohn's hypothesis of an TPD over the Arctic Ocean. Nansen proposed that enormous masses of drifting ice are exported by currents and winds, from the marginal seas of north-east Siberia, which cross the polar basin and then move into the Greenland Sea. Based on the drift of the Jeannette and Fram, Nansen estimated the duration of the TPD to be five to six years. During this time the ice cover undergoes significant changes. The ice floes, under the influence of wind and currents, constantly collide, break and ridge, piling up to form high ice billows.

Nansen's work has for many decades determined the strategy and tactics of polar research in the central Arctic. Nansen was the first to propose and implement the idea of a drifting scientific station in the ice (Nansen, 1903).

Intense scientific exploration, both in coastal seas and ocean areas in the Arctic continued at the beginning of the 20th century. A number of well-known expeditions contributed greatly to Arctic science.

In this connection we would like to mention the Russian explorer Alexander Kolchak who was active in the organization of a number of Russian expeditions at the beginning of

20th century. Using the drifts of the vessels Jeanette and Fram, Kolchak described the movement of ice under the influence of the winds and currents in the seas of the Siberian shelf. He also suggested a scheme for the movement of the Arctic pack ice for the whole polar basin. He specified that the influence of wind causes the Arctic pack ice to move westward along the edge of the Asian continent in the Siberian arctic region (Kolchak, 1909).

Nansen's ideas were put into practice in 1937 by the Soviet Union's Chief Office for the Northern Sea Route (now the Arctic and Antarctic Research Institute, St. Petersburg, Russia). On May 21, 1937 an airplane made the first landing near the North Pole on an ice floe at 89°43'N, and the first scientific drifting station North Pole-1 (NP-1) was set up (Figure 1).

The drift of NP-1 in 1937-1938 showed that ice in the western Arctic drifted in the same direction as 40 years earlier (Fram drift). That is to say from the North Pole southward to the Greenland Sea. But what is the pattern of the SID in the eastern Arctic? The answer to this question was obtained in the same years based on data acquired during the 1937 navigation season, when more than 20 Soviet ships, including ice-breakers, were frozen into the ice along the Northern Sea Route. The most remarkable was the drift of the ship Sedov that began in October 1937. The ship was gripped by ice in the Laptev Sea, south-west off the New Siberian Islands, and transformed into a scientific polar station. At the beginning, it was assumed that the drift would be along the westward route (Nansen route). In fact sea currents carried the ship to the east, and Sedov crossed the 150°E longitude and reached the region north-west of De Long islands. The direction of drift then changed, and Sedov began to drift close to the course that the Fram had drifted earlier. Thus existence of the TPD as one of the main features of the SID in the Arctic Ocean, as suggested by Mohn, Nansen and Kolchak, was confirmed. A second Russian drifting station, NP-2, was established in 1950, and from that time up to the present, manned drifting stations have continued to make observations, with the exception of a 12-year break from 1991-2003 (Figure 1). In some years three drifting stations in different areas of the Arctic Basin were in operation simultaneously. In September 2007 the station NP-35 started operations at the point 81°26'N, 103°30'E. The drifts of the NP-2 and NP-3 (1954–57) and later NP-22 (see Figure 1, red line) demonstrated the existence of a second system of the SID in the Arctic Ocean, with a circular clockwise drift. Similar data were obtained by the American T-3 station (also known as Fletcher's Ice Island). T3 set up on an ice island in March 1952, functioned intermittently until 1973 and made three circuits in the Beaufort Gyre before exiting the Arctic Ocean through the Fram Strait via TPD.

Following Nansen's idea of investigating the Arctic Ocean using drifting ice, the International Arctic Buoy Program (IABP) has since 1979 deployed buoys on ice to be tracked by satellite (http://iabp.apl.washington.edu). The objective of the IABP is to establish and maintain a network of data buoys in the Arctic Ocean to provide meteorological, sea ice and oceanographic data for real-time operational requirements and research purposes, including support to the World Climate Research Program (WCRP), the World Meteorological Organization (WMO) and World Weather Watch (WWW) Program.

Sea ice research using satellite remote sensing data became popular in the late 1970's and early 1980's. Seasat (Sea satellite) was launched in 1978 carrying various microwave sensors. Seasat sensors, such as a microwave scatterometer, microwave altimeter, microwave radiometer, and synthetic aperture radar have developed and improved over the years and have become indispensable oceanographic tools.

A number of geostationary meteorological satellites were put into orbit over the equator in the 1970s. In 1978 the Nimbus-7 with a Scanning Multichannel Microwave Radiometer (SMMR) was launched, and the Defense Meteorological Satellite Program (DMSP) -F8, -F11 and -F13 Special Sensor Microwave/Imager (SSM/I) followed in 1987. No data coverage is available for regions poleward of latitudes 84.5° N for SMMR and 87° N latitude for SSM/I, due to the inclination of the satellite orbits. SMMR data are acquired every other day, while SSM/I data are acquired daily. Satellite-observed data started to become available to scientists in the early 1980s. Fast and reliable observations of the ocean and the marine atmosphere by satellite-borne sensors were realized, and researchers are increasingly able to utilize this satellite-generated data. Satellite data are also important for continually updating models. It will take some time until the measurement technology and methodology are developed so that satellite ocean observation becomes a widely used tool for air-sea interaction research, and satellite ocean monitoring is established as an integral part of the earth environment observation system.

In 2009, the European Space Agency (ESA) will make another significant contribution to research in polar-regions with the launch of CryoSat-2, the agency's Earth Explorer ice mission. CryoSat-2 will monitor changes in the thickness of the polar ice sheets and floating sea ice. The observations made over the three-year lifetime of the mission will provide conclusive answers for the rates at which the ice cover is diminishing.

The International Council for Science (ICSU) and the World Meteorological Organization (WMO) have established the International Polar Year (IPY), a large scientific programme focused on the Arctic and the Antarctic from March 2007 to March 2009 (http://classic.ipy.org).

As a part of the IPY a scientific expedition on board the polar schooner Tara has been organized as a major part of the European scientific programme "Developing Arctic Modelling and Observing Capabilities for Long-term Environmental Studies (DAMOCLES)." This extensive programme gathers more than 45 laboratories with the aim of developing a long lasting device which can observe the ice, the sub-glacial ocean (temperature, salinity, currents), the atmosphere, and energy fluxes between air, ice and water (http://www.damocles-eu.org/).

Tara started drifting in the Laptev Sea in September 2006. During the first year she has covered 3 400 km with an average speed of 9,3 kilometres per day. During her long journey caught in the ice of the TPD Tara reached 88°32' North, just 160 km from the geographic North Pole on the 28th of May 2007. Tara is thus the icebound ship that has reached the most northerly position. In January 2008 Tara exited the Arctic Ocean through the Fram Strait and fulfilled her mission. The acceleration of the TPD might be partly responsible for the drastic decrease of the Arctic sea ice cover at the end of summer 2007 (Gascard, 2007).

Data

The following data sets were used in this study:

1. Ice motion vectors computed from Advanced Very High Resolution Radiometer (AVHRR), Scanning Multichannel Microwave Radiometer (SMMR), Special Sensor

Microwave/Imager (SSM/I), and International Arctic Buoy Programme (IABP) buoy data. Monthly mean gridded fields combine data from all sensors from November 1978 through December 2006 with a spatial resolution of 25x25 km (Fowler, 2003).
2. The monthly mean sea ice concentration data set generated from brightness temperature data derived from the Nimbus-7 Scanning Multichannel Microwave Radiometer (SMMR) (October 1978 to August 1987) and Defense Meteorological Satellite Program (DMSP) Special Sensor Microwave/Imager (SSM/I) (1987 to present) radiances at a grid cell size of 25 x 25 km (Cavalieri et al., 1996).

Both data sets are distributed by the National Snow and Ice Data Center Centre (NSIDC), Boulder, Colorado USA (http://nsidc.org).

We also used the monthly mean sea level atmospheric pressure (SLP) fields gridded with spatial resolution 5 degrees for the period 1947-2006 and monthly mean wind vectors derived from the National Centre for Environmental Prediction (NCEP) reanalysis obtained from the National Centre for Atmospheric Research (NCAR), NCEP/NCAR Reanalysis project data sets (http://dss.ucar.edu). All these data were interpolated into the Northern Hemisphere EASE-Grid.

Seasonal Variability of Sea Ice Drift

The mean SID and SLP fields are calculated using data from 1979 to 2005. The seasonal variability is shown in Figure 2. In winter time (December-March) the Beaufort high SLP and the low SLP areas over the Bering, Norwegian and Barents seas drive the anticyclonic SID in the Beaufort Gyre and the TPD from the northern parts of Chukchi and East-Siberian seas and the Laptev Sea, across the Arctic Ocean to the Fram Strait. In December the Beaufort Gyre and TPD are at a maximum. At this time maximum SID velocities are observed at the southern periphery of the Beaufort Gyre along the Alaska coast, in the TPD, in the Baffin Bay and the northern part of the Barents Sea, and in the Fram Strait. Near the North Pole, the TPD forks into two branches. The first branch joins the north-west periphery of the Beaufort Gyre and the ice drifts southwards towards the Alaska coast. The second branch is directed towards the Fram Strait. SID from the Kara Sea is mostly northward. At 80°N it divides into two branches. In the eastern branch, sea ice passes through the strait between Franz Josef Land and Severnya Zemlya islands, and comes into contact with the southern border of the TPD. Sea ice in the western branch enters the Barents Sea through the strait between Franz Josef Land and Novaya Zemlya. SID in the Laptev Sea is north-eastward and also joins the TPD near the continental slope. On the shelf of the East Siberian Sea in December and January, SID is eastwards, but velocities are much smaller than in the other marginal seas. SID in the East Siberian Sea has a similar structure for all the winter months but turns north-westward in February and March, thus contributing to the TPD. The SID fields in spring (April, May) are very similar to those seen in the winter, but velocities are lower. The summer months (June-September) are accompanied by a decrease of the SLP over the Arctic Ocean, Siberia, Canada, Alaska and Greenland and increased SLP over the Norwegian and Barents seas. This leads to decreased SLP gradients and lower wind and SID velocities. In the northern part of the Laptev Sea in June and July a mesoscale cyclonic eddy is observed.

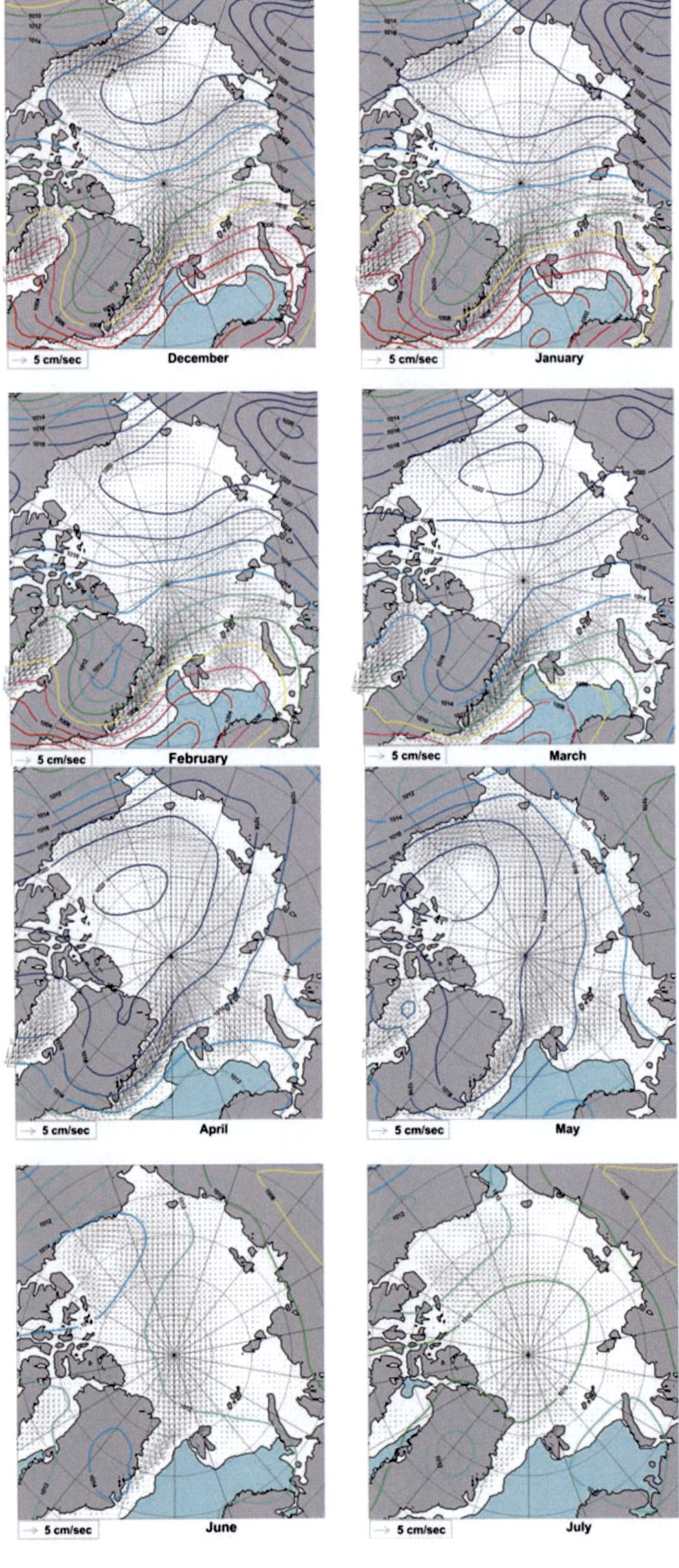
5 cm/sec
December
5 cm/sec
January
5 cm/sec
February
5 cm/sec
March
5 cm/sec
April
5 cm/sec
May
5 cm/sec
June
5 cm/sec
July

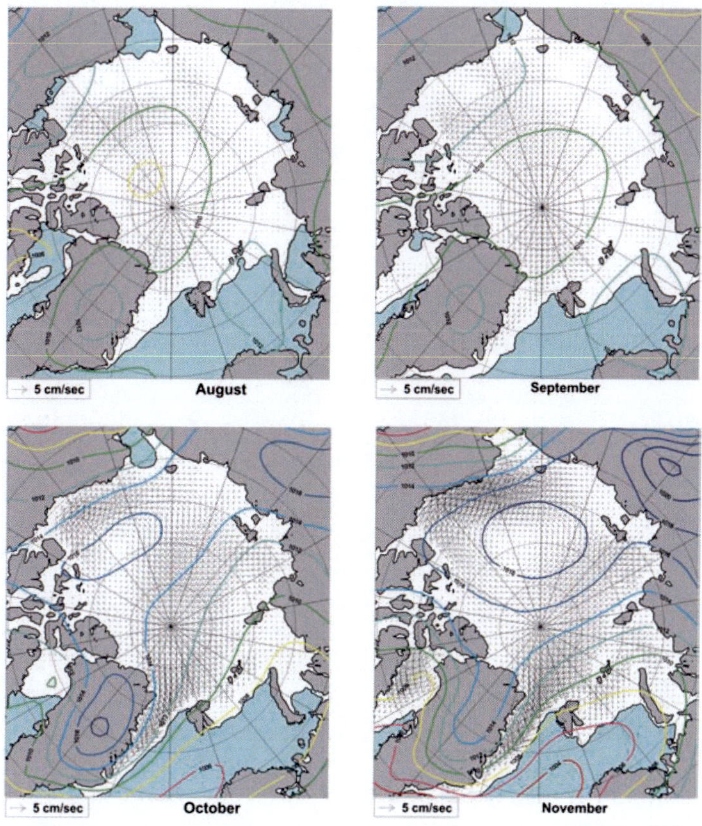

Figure 2. Annual variability of the mean sea ice drift (SID) and sea level atmospheric pressure (SLP).

The Beaufort Gyre becomes weaker in June and July, and by August it has completely disappeared. In August a cyclonic SID is formed in the Arctic Ocean, centred near 85°N, 160°E. In September the Beaufort Gyre is re-established, and the cyclonic eddy in the northern part of the Laptev Sea is disappearing. SID velocities increase during October and November, with SLP and SID fields resembling the winter structures.

The location of the centre of the Beaufort Gyre changes over the year (Figure 3). From January to July the centre of Beaufort Gyre rotates anticyclonically, following a cyclonic trajectory from September to January.

Over short time scales SID can be approximated by the simple rule known as 'isobaric SID,' which was formulated by Zubov (1943) and has been used in many studies (e.g., Thorndike and Colony, 1982; Colony and Thorndike, 1984; Rigor, Wallace and Colony, 2002). According this rule sea ice drifts along instantaneous isobars with a velocity proportional to the SLP gradient across isobars. SID across isobars occurs in areas where the influence of the ocean currents is significant. Using IABP data, Rigor, Wallace and Colony (2002) analyzed winter and summer SID climatology and reported roughly equal contributions from the wind and ocean current in driving SID. Earlier estimations of the influence of ocean currents on SID, made by Buynitskiy (1951) and Vorobyev and Gudkovich (1976), show it to be around 60% on average. The influence of currents is strongly seasonal. Figure 2 shows that during the winter season and in April and May the SID in the main part of the Arctic Ocean is directed along isobars, and the role of ocean currents is small.

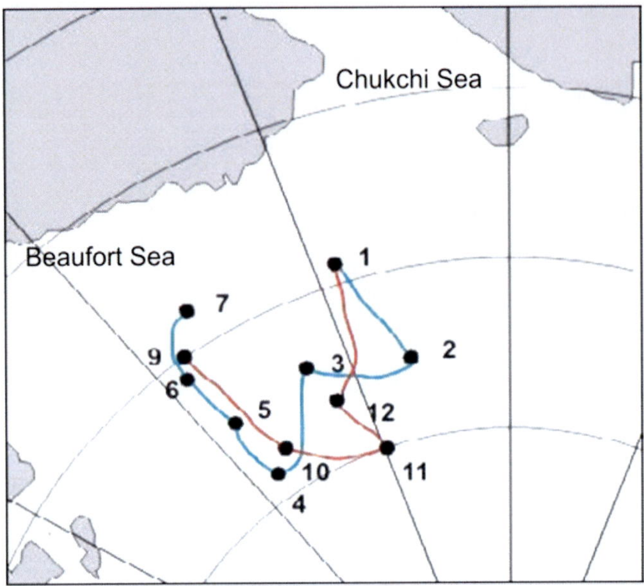

Figure 3. The location of the centre of the Beaufort Gyre for each month of the year.

At this time of year it is only in the Barents Sea, the Baffin Bay and in the area to the north of Fram Strait that the SID crosses isobars due to the currents' influence. In the summer (June-September), ice melting and water warming cause a dramatic increase in density gradients, especially in marginal seas. This leads to an increase in density-driven currents and their contribution to ice circulation becomes larger. In June isobaric SID is observed only in the Beaufort Sea. In other areas the angle between SID directions and isobars are between 30° and 90°. In July isobaric SID disappears completely. In August, when low SLP occurs in the central part of the Arctic Ocean, SID returns to an isobaric configuration everywhere except in the Beaufort Sea and the area north of the Fram Strait. In September, when the anticyclonic Beaufort Gyre reappears, the wind-driven component of SID prevails in all parts of the Arctic Ocean. Thus the combined statistical analysis of the sea ice velocities together with SLP and wind fields over the Arctic Ocean shows that the seasonally changing local wind is the most important factor determining the annual cycle of sea ice velocities.

Mirroring seasonal changes of the wind in the Arctic, SID velocities reach a maximum in December and are at a minimum in June (Figure 4). In contrast to the present study, Frolov et al. (2005) reported minimum SID velocities in March and maximum velocities from August to October. This disagreement possibly occurs because Frolov et al. (2005) used data from Russian North Pole (NP) Stations, which operate mostly in the central part of the Arctic Ocean. The annual variability of SID velocities is very different in different parts of the Arctic Ocean. Figure 5 shows the spatial distribution of the amplitude of SID velocities annual cycle and also the time (month) of the minimum and maximum velocities in the Arctic Ocean. In the central part of the Arctic Ocean maximum SID velocities are observed in summer (August, September) and minimum velocities are observed during the winter (December-March).

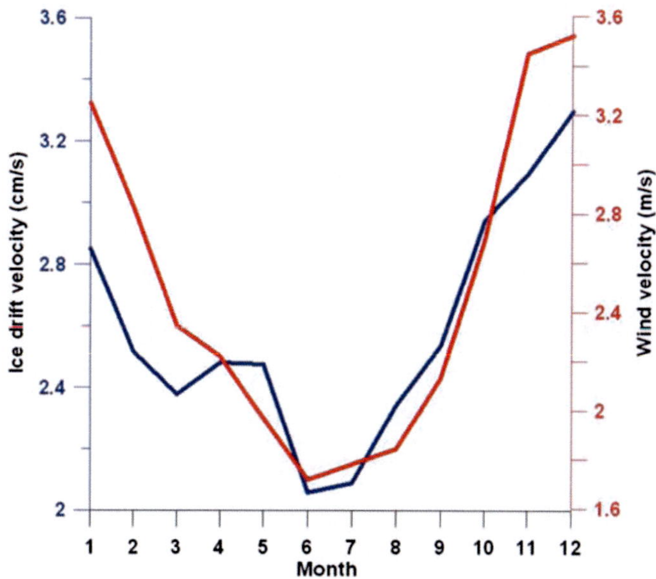

Figure 4. The seasonal variability of sea ice drift (SID) velocity (blue) and wind velocity (red) averaged for the entire Arctic Ocean.

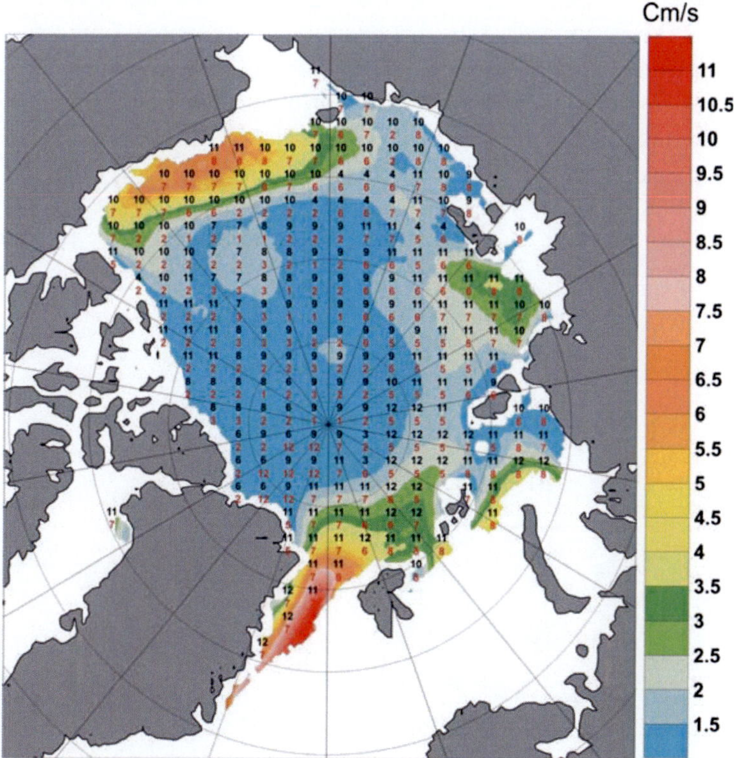

Figure 5. The spatial distribution of the amplitude of the annual cycle of sea ice drift (SID) velocities. The months with minimum velocities (red) and maximum velocities (black) are also shown.

The opposite occurs in the Fram Strait, the Beaufort Sea and the marginal Arctic seas. In all these areas SID velocities reach maximum values in the late autumn or beginning of winter (October, December). Minimum SID velocities here are observed in the summer (June-August).

The average amplitude of the annual cycle of the SID velocities in the Arctic Ocean is of the order of 2-3 cm s^{-1}. In the central part of the Arctic Ocean this amplitude is not more than 2 cm s^{-1}. The maximum amplitudes are found in the Fram Strait (9-10 cm s^{-1}), Beaufort Gyre (6-7 cm s^{-1}) and the northern part of Barents Sea (5-6 cm s^{-1}).

SID velocities also vary widely throughout the Arctic Ocean. Annual mean SID velocities for the period 1979-2005 (Figure 6a) range from 1 to 3 cm s^{-1} in the central part of the Arctic Ocean, from 3 to 5 cm s^{-1} in the Beaufort and Barents seas and Baffin Bay, and from 5 to 15 cm s^{-1} in the Fram Strait. Figure 6b shows maximum values of the monthly mean SID velocities observed in period from 1979-2005. Maximum SID velocities are 5-6 times higher than annual mean. There are four areas with high velocities: the Beaufort Sea (15-20 cm s^{-1}), the Barents Sea and Baffin Bay (14-16 cm s^{-1}) and the Fram Strait (25-30 cm s^{-1}).

Sea ice is exported from almost all areas of the Arctic Ocean, through the Fram Strait and into the Greenland Sea. Probable travel times for ice from different parts of the Arctic Ocean to the Fram Strait were estimated by Pavlov; Pavlova and Korsnes (2004). The probability of sea ice export from the area just north of the Fram Strait is the highest (88-90%), with a travel time less than a year. Sea ice drifting from shelf areas of the Kara and Laptev seas and from the western part of the East Siberian Sea also has rather short travel times (3-6 years, on average), whereas sea ice from the Beaufort Sea region reaches the Fram Strait in 6-7 years on average. The maximum travel time (10-16 yr) for sea ice to reach the region of the Fram Strait, with rather low probabilities (38-65%), is for sea ice from the areas north of the Canadian archipelago. These estimates agree well with similar estimates by Rigor, Wallace and Colony (2002) and Pfirman et al. (1997b) based on analyze of IABP data.

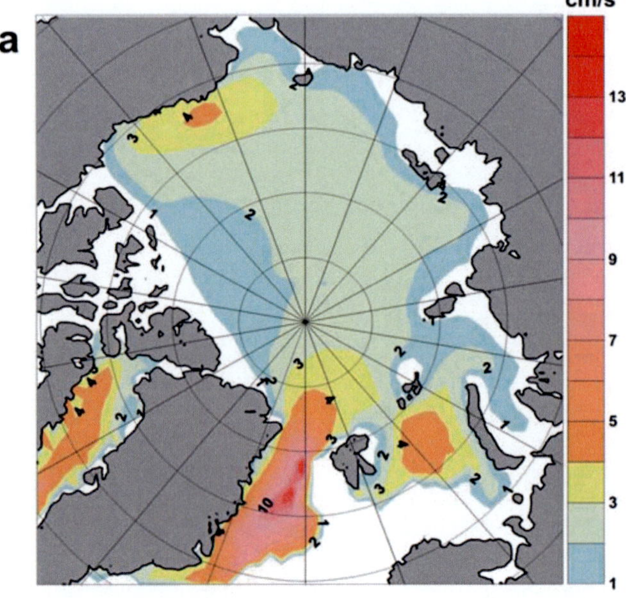

Figure 6. (Continues)

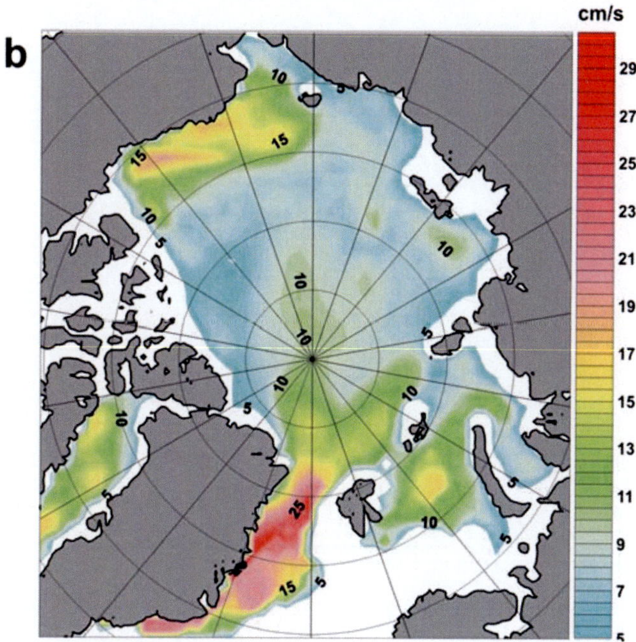

Figure 6. Annual mean (a) and maximum monthly mean (b) sea ice drift (SID) velocities for the period 1979-2005.

SEASONAL SEA ICE EXCHANGE BETWEEN THE ARCTIC BASIN AND ARCTIC SEAS

The mean sea ice flux has previously been estimated based on satellite observations and modelling results both for the Fram Strait (e.g., Kwok and Rothrock, 1999; Kwok, Cunningham and Pang, 2004; Pavlov, Pavlova and Korsnes, 2004; Hop et al., 2006; Hop and Pavlova, 2008) and the Barents Sea entrances, represented by Svalbard-Franz Josef Land and Franz Josef Land-Novaya Zemlya channels (e.g., Pavlov, Pavlova and Korsnes, 2004; Kwok, Maslowski and Laxon, 2005; Hop and Pavlova, 2008), and for the other main transects in the Arctic Ocean (Pavlov, Pavlova and Korsnes, 2004; Frolov et al., 2005).

Our current study represents an update of the previous estimates using 27 years (1979-2005) of satellite passive microwave data records of SIC and ice motion vectors. To estimate the ice area flux through the main straits in the Arctic Ocean we have applied the method described in Kwok and Rothrock (1999), but with one difference: in calculating the ice area flux in the Fram Strait Kwok and Rothrock assumed 100% concentration within the 15% ice edge, while we have used the full range (0.1-1.0) of the SIC (see also Pavlov, Pavlova and Korsnes, 2004).

The calculation was made for seven transects (Figure 7). A positive value implies a flux out of the Arctic Ocean. The sea ice flux has a strong seasonal variability (Figure 8). In the Fram Strait (Transect 1) the maximum export of sea ice to the Greenland Sea (about 80×10^3-90×10^3 $km^2 month^{-1}$) takes place during the winter from December through April. After May, the ice flux decreases sharply and in August reaches its minimum (about 8×10^3 $km^2 month^{-1}$).

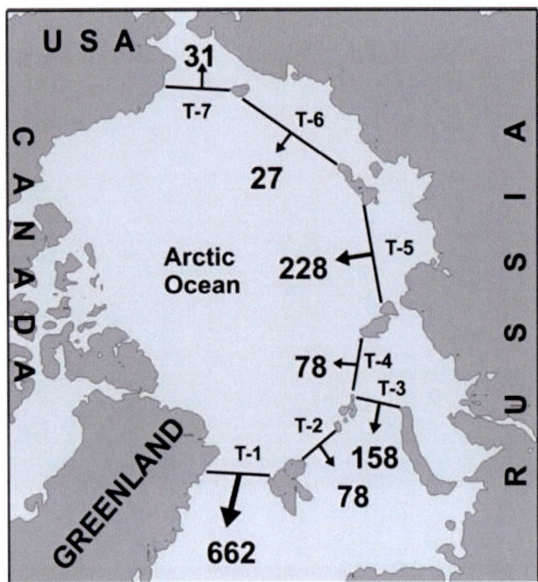

Figure 7. Simulation domain and positions of transects. Mean annual sea ice flux averaged for the period 1979-2005. Area flux (10^3km^2).

In the strait between Svalbard and Franz Josef Land (Transect 2) the ice flux has the same direction throughout the year (from the Polar Basin to the Barents Sea). The minimum value (about 690 km^2 month^{-1}) is in August and there are two maxima (about 10×10^3 km^2 month^{-1}), in April and December.

In the entrance between Franz Josef Land and Novaya Zemlya (Transect 3), the average ice flux is close to zero during the summer months, from July to September. During the other seasons of the year, ice flux is directed from the Kara Sea to the Barents Sea reaching a maximum (25×10^3 km^2 month^{-1}) in December.

In the straits between Franz Josef Land and Severnaya Zemlya (Transect 4), and Severnaya Zemlya and the Novosibirskie islands (Transect 5) the ice fluxes are directed into the Polar Basin over the whole year, except during three summer months (June-August) for Transect 4. The ice flux in November-February from the Kara Sea is between 11 x 10^3 and 17.5 x 10^3 km^{2}month^{-1}. From the Laptev Sea the ice flux is between 29 x 10^3 and 49 x 10^3 km^{2}month^{-1}. In the summer there is a small ice flux to the Kara Sea through the strait between Franz Josef Land and Severnaya Zemlya (between 1.0 x 10^3 and 1.7 x 10^3 km^{2}month^{-1}).

The opposite situation is seen on the northern boundaries of the East Siberian and Chukchi seas. The ice flux through the transect between the Novosibirskie islands and Wrangel Island (Transect 6) has a very complicated character. Ice flux is directed into the Polar Basin in the period from February to October (maximum in May of about 14 x 10^3 km^2) with the exception of August, when ice flows in the opposite direction. From November-January there is also an inflow to the East Siberian Sea, with a maximum in November (about 14 x 10^3 km^2). Through the transect between Wrangel Island and Icy Cape (Transect 7) sea ice flows from the Chukchi Sea to the Polar Basin from March to September with a maximum in May (11 x 10^3 km^2). In the other months the ice flux changes direction and the import of sea ice from the Polar Basin during this period reaches a maximum of 24 x 10^3 km^2 in December.

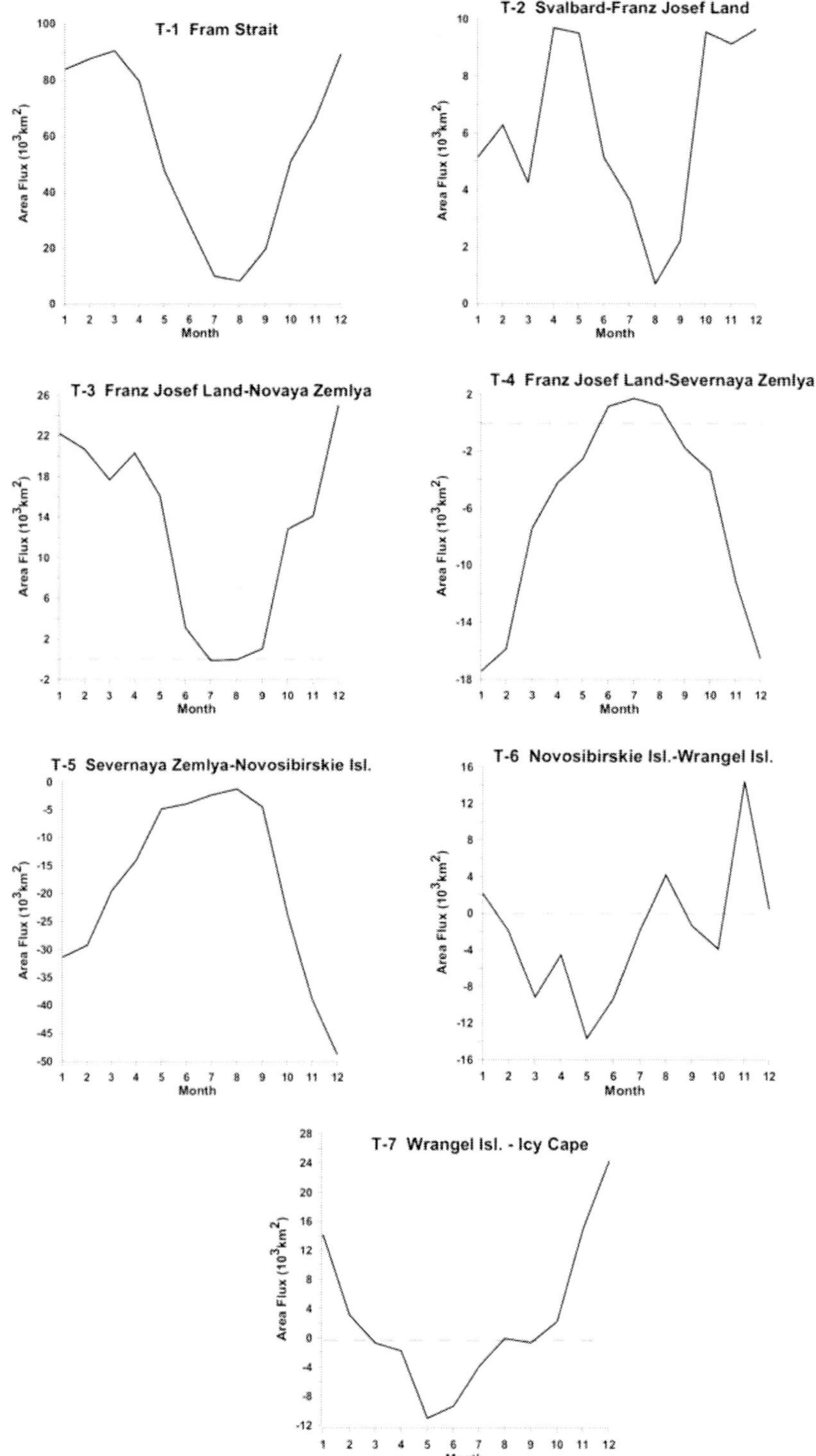

Figure 8. Seasonal variability of the sea ice flux through the transects averaged for the period 1979-2005. A positive value implies a flux out of the Arctic Ocean. The locations of the numbered transects are shown in Figure 7.

INTER-ANNUAL VARIABILITY OF THE SEA ICE DRIFT

The variability of the monthly mean SID velocities averaged for the entire Arctic Ocean is shown in Figure 9.

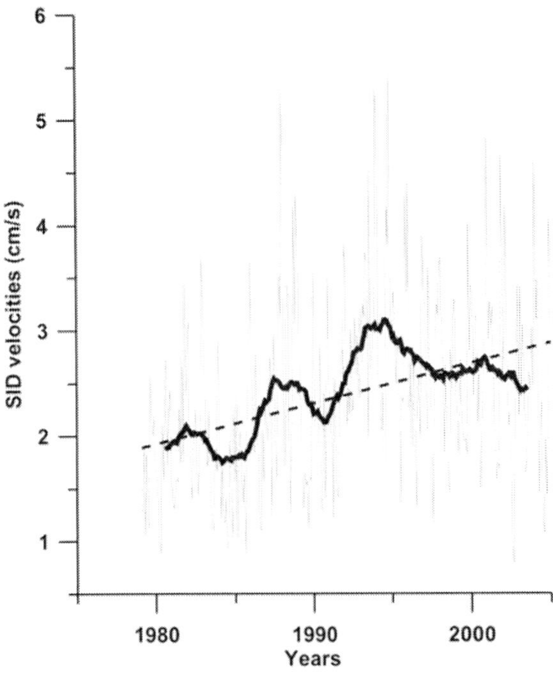

Figure 9. Variability of the monthly mean sea ice drift (SID) velocities averaged for the entire Arctic Ocean - thin line; 3-year running mean – thick line, linear trend - dashed thick line.

The main feature of the SID velocities variability is a clear positive trend. Against a background of this positive trend, fluctuations with different periods are also presented here. A Fast Fourier Transform (FFT) analysis of the spectrum of SID velocities gives three significant peaks with periods of 12.2 months, 31.6 months (2.6 years) and 63.2 months (5.3 years) having (Figure 10a). The first peak corresponds to the period of 12 months related with annual cycle of the SID velocities described above. The peak corresponding to a period of 63.2 months (5.3 years) has almost the same significance as the annual fluctuations (Figure 10a). A similar spectrum is also found in the variability of the wind velocities (Figure 10b). A spectral analysis of the wind velocity variability also gives three significant peaks with periods of 11.8 months, 29.2 months (2.4 years) and 75.8 months (6.3 years). Fluctuations with nearly the same periods are also found in the variability of SIC anomalies in the Arctic Ocean.

Figure 11 shows the 3-year running mean variability of the three parameters: anomalies of SID velocity, wind velocity and SIC, where annual cycles and linear trends are excluded. There is a good positive correlation not only between SID and wind velocities, which is the main driving force, but also between SID and SIC. Within the observation period each of these parameters oscillates with period 5-6 years reaching maxima in 1981-1983, 1987-1989 and 1993-1995 and minima in 1984-1986, 1990-1992 and 1996-1998.

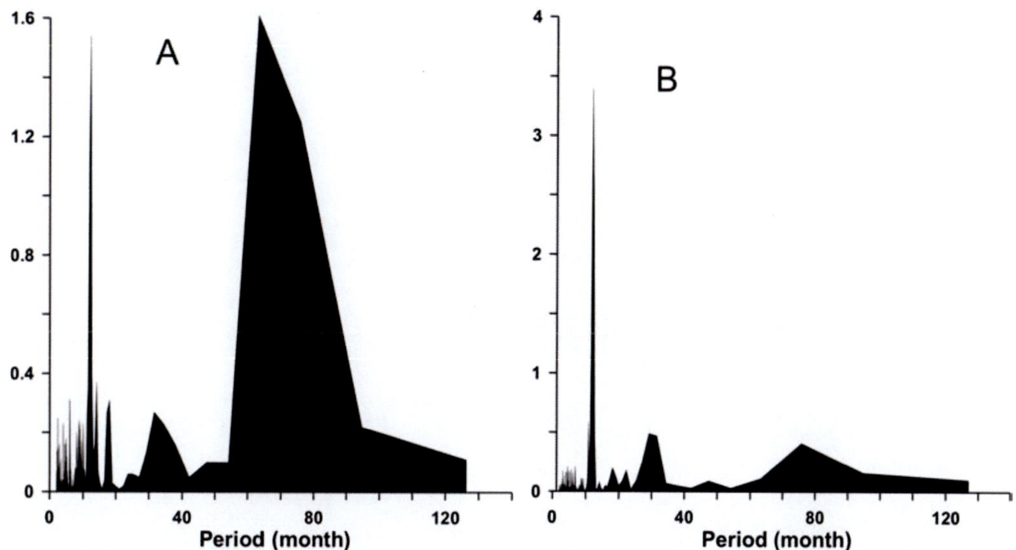

Figure 10. The spectral Fast Fourier Transform (FFT) analysis of the sea ice drift (SID) velocities (a) and wind velocity (b).

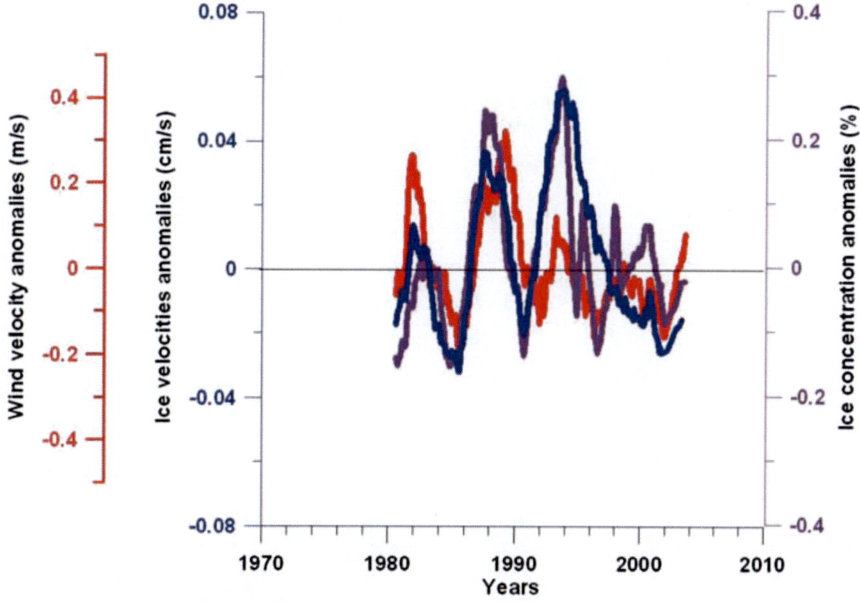

Figure 11. 3-year running means of the anomaly of sea ice drift (SID) velocity (red), anomaly of wind velocity (blue) and anomaly of sea ice concentration (SIC) (violet).

Figure 12 shows the large differences in the winter patterns of the atmospheric pressure field and ice circulation that correspond to maximum and minimum SID velocities. During the maximum positive SID velocities, wind velocities and SIC anomalies, the Icelandic Low is very well developed, with a field of low SLP extending from Iceland and the Norwegian Sea far to the northeast, covering large areas of the Barents and Kara seas, and a large area of the Western Siberia. In the central part of the Arctic Ocean a high SLP anomaly is forming.

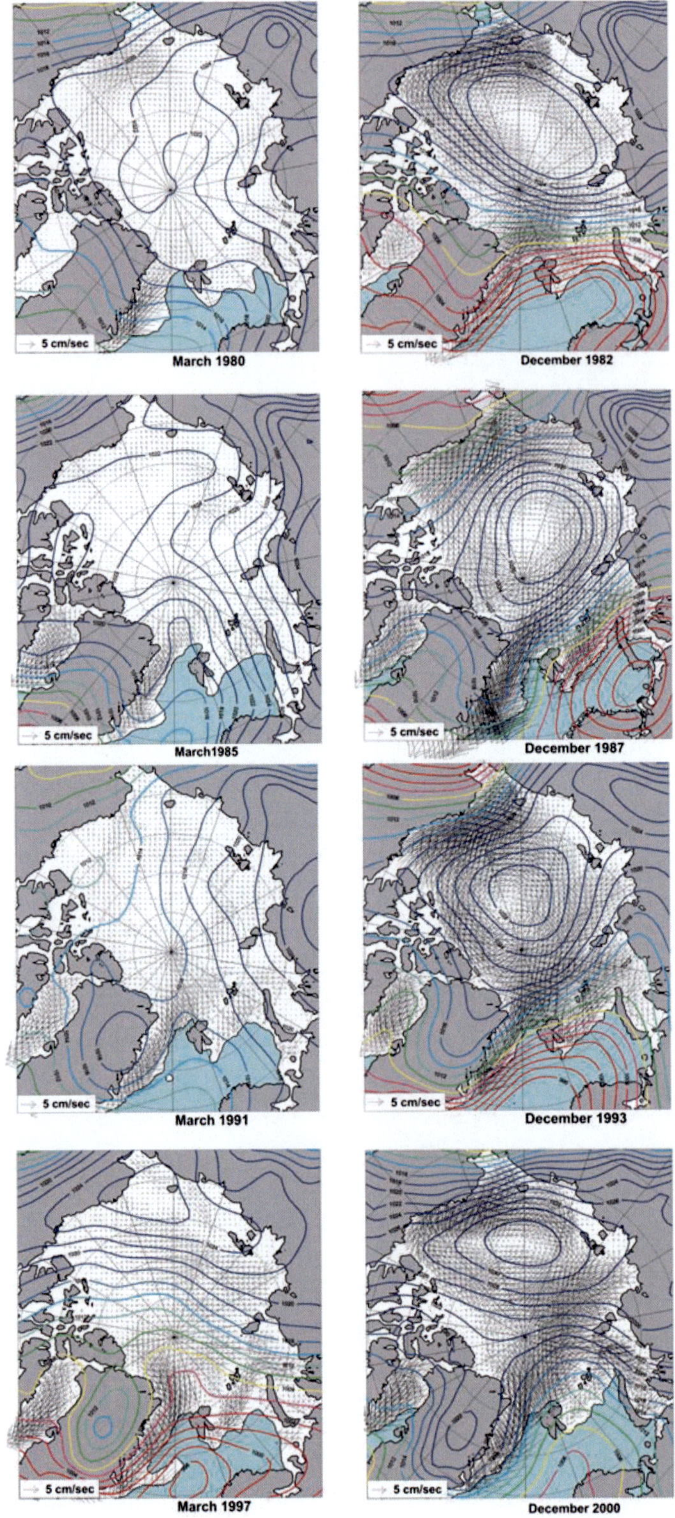

Figure 12. Sea ice drift (SID) and sea level atmospheric pressure (SLP) patterns for years with minimum (left panels) and maximum (right panels) SID velocities.

There are strong positive SLP gradients between coastal areas (Siberia, Canada and Alaska) and the central Arctic Ocean. This leads to an intensification of the anticyclonic rotation and increased southeast winds over the East Siberian and Laptev seas and the Siberian continental slope. Thus in years with maximum SID velocities, the ice circulation patterns are characterized by a well-developed Beaufort Gyre and a strong TPD.

In years with minimum SID velocities, Beaufort Gyre and TPD are destroyed completely, and along the Siberian shelf the prevailing direction of ice motion is eastward. Such fluctuations in SID have been widely discussed in the literature. Using a simple 2-D barotropic model to simulate circulation of water and ice circulation in the Arctic Ocean, Proshutinsky and Johnson (1997) have corroborated the existence of the two regimes of SID, proposed by Gudkovich (1961). Proshutinsky and Johnson (1997) found that shifts between cyclonic and anticyclonic flow occur every 5-8 years with a period of oscillation of about 10-15 years. By contrast our results agree with Karklin (1977) and many other Russian investigations referred to by Proshutinsky and Johnson (1997). They show shifts between cyclonic and anticyclonic regime of the SID occurring every 2-3 years, with a period of oscillation of 5-6 years. This contradiction possibly occurs because we use only 27 years of records for our analysis, which is not a sufficiently long period to observe oscillations with periods of 10-15 years.

In any case, as shown by many authors (e.g., Walsh, Chapman and Shy, 1996; Kwok, 2000; Hilmer and Jung, 2000; Zhang, Rothrock and Steele, 2000; Rigor, Wallace and Colony, 2002), the low frequency periodicity of the SID velocities is connected to the reorganization of the atmospheric circulation of the Arctic Ocean. The periods of the most significant oscillations of the main indexes of the atmospheric circulation in the period of 1979-2005 are close to the periods of SID, wind and SIC variability. The North Atlantic Oscillation (NAO) index has periods of 12.3 months, 23.1 months (1.9 years), 63.9 months (5.3 years) and 95.8 months (8 years); the Arctic Oscillation (AO) index has periods of 12.4 months, 31.9 months (2.7 years), 74.6 months (6.2 years) and 111.8 months (9.3 years); and the North Pacific Oscillation (NPO) index has periods of 12.0 months, 29.2 months (2.4 years), 67.1 months (5.6 years) and 95.8 months (8 years).

Figure 13 shows the spatial distribution of correlation coefficient between SID velocity, averaged for the entire Arctic Ocean, and SLP over the Arctic. Mainly negative correlation values are seen over the north Pacific and north Atlantic oceanic areas, while positive values of the correlation coefficient are seen over the land areas of Eurasia and North America. There are two areas with maximum values of the negative correlation coefficient: in the north Pacific-area near Alaska coast (maximum correlation coefficient R=-0.47), and in the north Atlantic- the Barents Sea (maximum correlation coefficient R=-0.53). Both correlation coefficients have 95% confidence limit. These maximum correlations are connected with the NPO and NAO correspondingly.

Maximum values of positive correlation coefficient (R=0.35-0.40) are seen in the mid-latitudes of Eurasia, which can be explained by the AO. This agrees with estimations by Rigor, Wallace and Colony (2002). They have found that the correlation between sea ice motion and AO during the winter time ranges from 0.35 near Fram Strait to 0.40 in the Chukchi and Beaufort seas. During the summer the correlation is near 0.30 throughout most of the basin. Thus the NAO, AO and NPO have approximately equal contributions to the averaged SID velocity variability for the entire the Arctic Ocean.

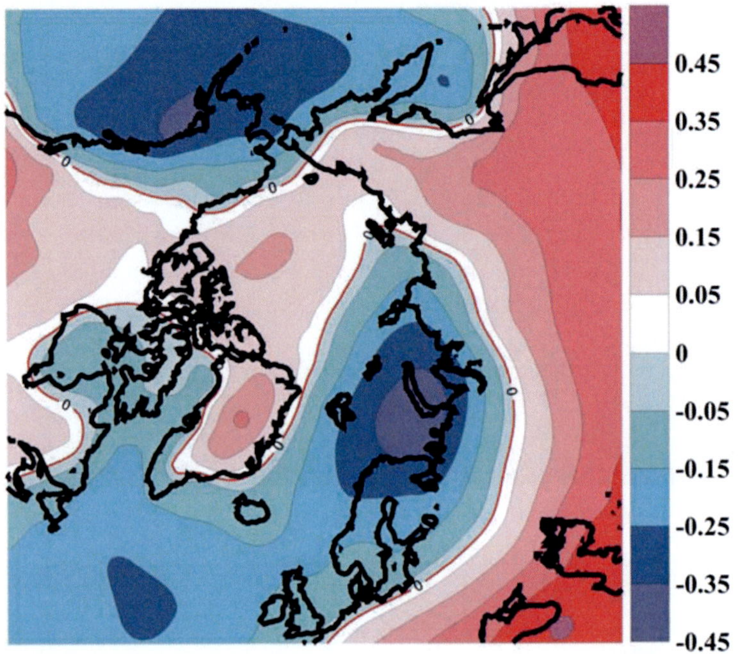

Figure 13. Spatial distribution of the correlation coefficient between sea ice drift (SID) velocity, averaged for the entire Arctic Ocean, and sea level atmospheric pressure (SLP) over the Arctic.

Regionally the NPO contributes more to variability of the Beaufort Gyre SID velocities; the NAO has the greatest effect on variability of SID velocities in the Fram Strait, and Barents and Kara seas, and the AO is mostly responsible for variability of SID velocities in the TPD and southern periphery of the Beaufort Gyre.

It is impossible to explain the well expressed positive trend in the SID velocity variability as a result of changes in the wind as the main forcing, because the trend of the average wind velocity over the ice in the period of 1979-2005 is negative (Figure 14). The spatial distribution of the SID velocity linear trend is shown in Figure 15. Trends of the monthly mean ice drift velocities are positive almost everywhere in the Arctic Ocean. In the Baffin Bay, Fram Strait and Barents Sea regions sea ice velocities have dramatically increased during the last two decades. Trends in these areas reached 0.15-0.20 cm s^{-1} per year. Analysis of the trends of the inter-annual sea ice velocities variability in the different seasons shows that maximum values of the linear trend appear in winter time (October-April) and minima in summer time (June-August). In August the trend is slightly negative (Figure 16). It is well known that the wind component of ice floes velocity increases with decreasing SIC and ice thickness (Lepparanta, 2005; Frolov et al., 2005).

Figure 17 shows the spatial distribution of the SIC trend, and we can see that areas with maximum negative SIC trends (Beaufort Sea, Siberian seas, Barents Sea, Fram Strait and Baffin Bay) correspond well with the areas of maximum positive trends of the SID velocities (see Figure 15). Rothrock, Yu and Maykut (1999) and Rothrock and Zhang (2005) also reported a substantial decrease of the sea ice thickness in the Beaufort Sea and Eastern regions of the Arctic Ocean.

A decrease in the SIC and ice thickness is observing during the last three decades, and this appears to be the main cause of the positive trend in SID velocities.

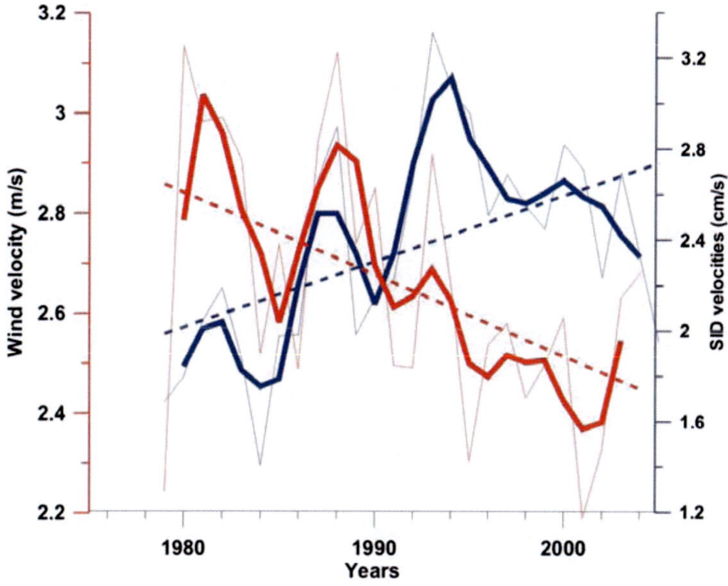

Figure 14. Inter-annual variability of the annual mean sea ice drift (SID) velocity, averaged for the entire Arctic Ocean (thin blue line), and wind velocity, averaged over the ice (thin red line). Thick lines are the 3-year running means of SID velocity (blue) and wind velocity (red).

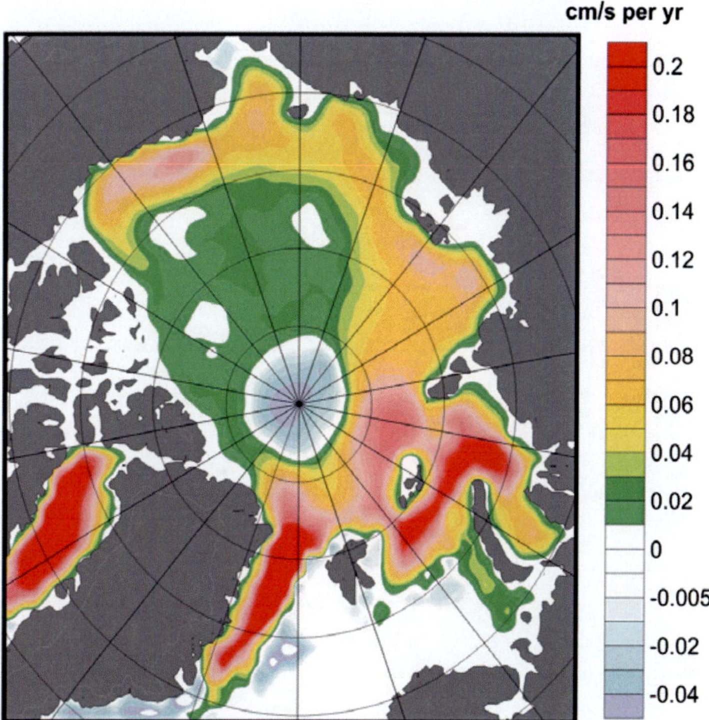

Figure 15. Spatial distribution of the sea ice drift (SID) velocity linear trend.

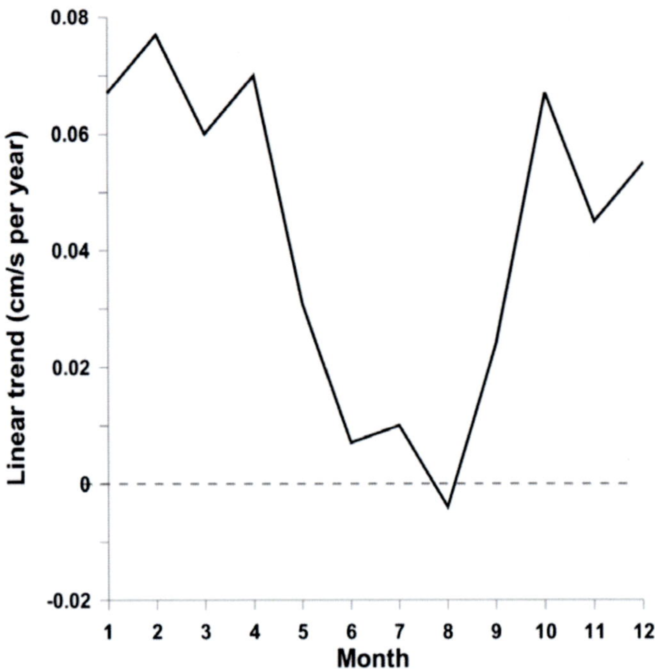

Figure 16. Seasonal variability of linear trend of the sea ice drift (SID) velocity.

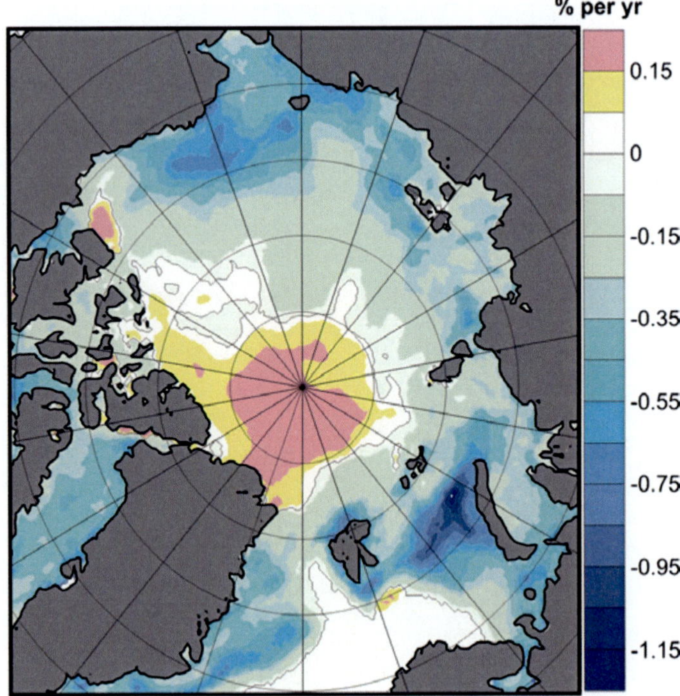

Figure 17. Spatial distribution of the sea ice concentration (SIC) linear trend.

INTER-ANNUAL SEA ICE EXCHANGE BETWEEN THE ARCTIC BASIN AND ARCTIC SEAS

The character of the inter-annual variability of ice exchange between the marginal Arctic seas and Arctic Basin and in the Fram Strait in general is similar to the variability of SID velocities averaged for the whole Arctic Ocean.

Mean annual climatic ice fluxes over the 27-year simulated record (1979-2005) are presented in Figure 7. The average sea ice export from the Arctic Ocean through Fram Strait (Transect 1) is about 662 x 10^3 km^2 y^{-1}, the average winter flux (October-May) is 598 x 10^3 km^2 y^{-1}. The average export through the strait between Svalbard and Franz Josef Land (Transect 2) is 78 x 10^3 km^2 y^{-1}, average winter flux is 65 x 10^3 km^2 y^{-1}. The summer months (June-September) contribute about 11% and 16% to the annual ice area exported through Transects 1 and 2 respectively. An estimate of the ice flux from the Kara Sea to the Barents Sea through the strait between Franz Josef Land and Novaya Zemlya (Transect 3) gives 158 x 10^3 km^2 y^{-1}, with an average winter flux of 154 x 10^3 km^2 y^{-1}. The summer months contribute only about 2% to the ice flux from the Kara Sea because of significant year to year variations in the ice flux direction across this transect in summer.

In contrast to Transects 1-3, the average sea ice fluxes over the same period through Transects 4-6 are directed from the Kara, Laptev and East Siberian seas into the Arctic Ocean, and have been estimated to be (see Figure 7): 78 x 10^3 km^2 y^{-1} through the straits between Franz Josef Land and Severnaya Zemlya (average winter flux is 81 x 10^3 km^2 y^{-1}), 228 x 10^3 km^2 y^{-1} between Severnaya Zemlya and Novosibirskie islands (average winter flux is 216 x 10^3 km^2 y^{-1}) and 27 x 10^3 km^2 y^{-1}between Novosibirskie islands and Wrangel Island (average winter flux about 18 x 10^3 km^2 y^{-1}). The summer months contribute only about 3% to the ice flux from the Arctic Ocean to the Kara Sea (Transect 4). For Transects 5 and 6 the summer ice flux has the same negative direction (into the Arctic Ocean) as in the winter and contributes about 5% and 33% respectively.

The average sea ice export from the Arctic Ocean through the strait between Wrangel Island and Icy Cape (Transect 7) is about 31 x 10^3 km^2 y^{-1}, and is strongly variable in direction from year to year. The average winter flux from the Arctic Ocean is 45 x 10^3 km^2 y^{-1}, while the summer ice flux is 14 x 10^3 km^2 y^{-1} in the opposite direction.

Figure 7 shows that the annual mean ice flux from all the marginal seas, except the Chukchi and Barents seas, is directed into the Polar Basin, giving an average export to the Polar Basin of 333 x 10^3 km^2 y^{-1}. The total import to the Greenland, Barents and Chukchi seas from the Polar Basin is 929 x 10^3 km^2 y^{-1}. The sea ice import (158 x 10^3 km^2 y^{-1}) from the Kara Sea to the Barents Sea through the strait between Franz Josef Land and Novaya Zemlya should also be noted.

Figure 18 shows the inter-annual variability of annual mean ice area fluxes over the 27 years from 1979 to 2005 in the seven straits discussed above. We have calculated the annual area flux as a flux for the period from October to September the next year. So, for example, in our calculation the annual mean ice flux for 1995 is the mean ice flux in the period October 1994 to September 1995.

In the Fram Strait (Transect 1) there are significant variations from year to year with a minimum of 282 x 10^3 km^2 in 1980 and a maximum of 1293 x 10^3 km^2 in 1995. Over the entire period the sea ice flux is directed out of the Arctic Ocean.

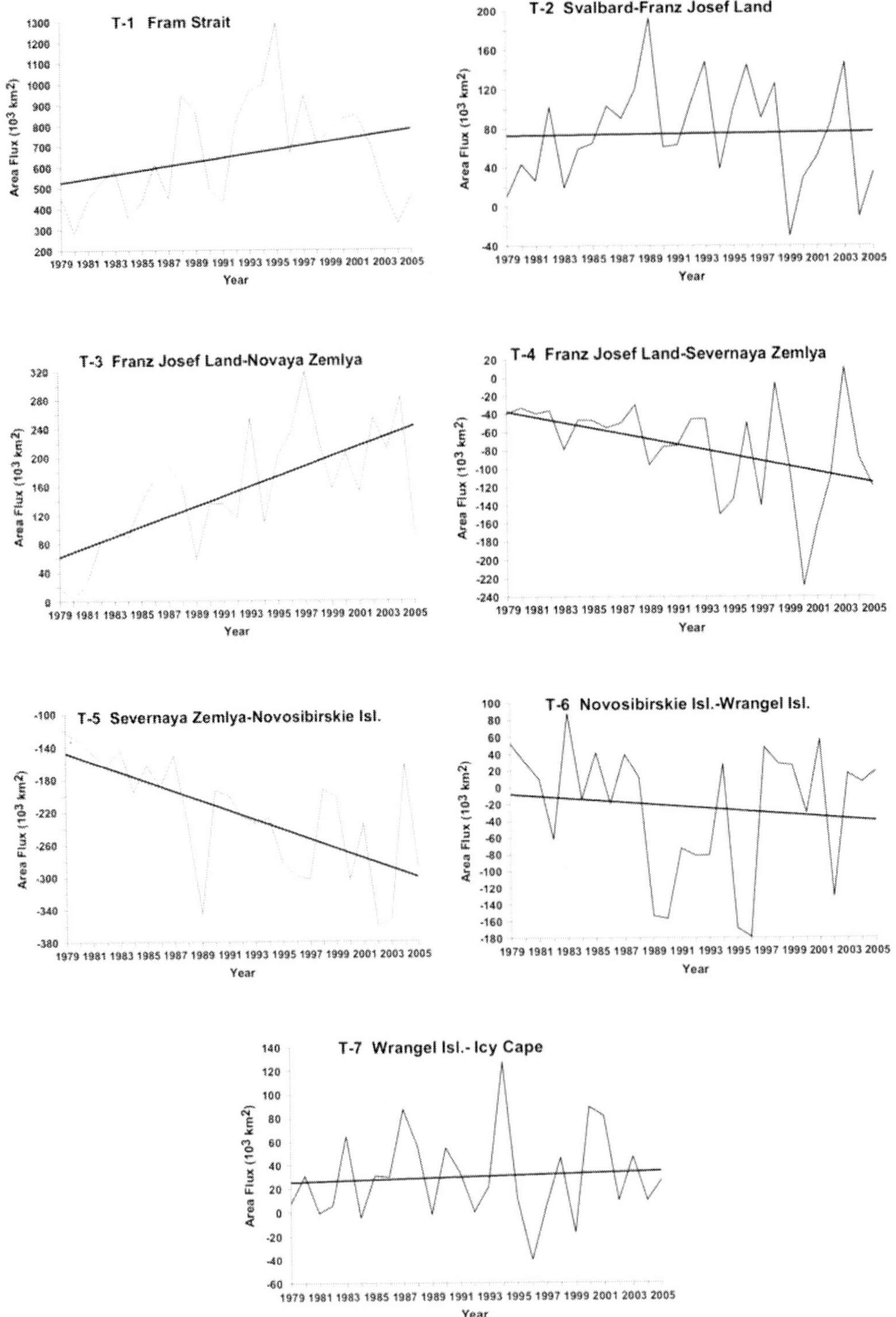

Figure 18. Inter-annual variability of annual mean ice flux through the transects for the period 1979-2005. A bold line shows the linear trend. A positive value implies a flux out of the Arctic Ocean. Locations of the numbered transects are shown in Figure 7.

The winter ice area flux ranges from a minimum of 257 x 10^3 km^2 in 1980 to a maximum of 1146 x 10^3 km^2 in 1995. The summer flux through this transect ranges from a minimum of 1.2 x 10^3 km^2 in 1984 to a maximum of 192 x 10^3 km^2 in 1994. For Transect 2 the maximum ice flux (193 x 10^3 km^2) was in 1989. In 1999 and 2004 the total area fluxes are -30 x 10^3 km^2 and -10 x 10^3 km^2 respectively, where the negative values indicate fluxes into the Arctic Ocean. In winter time the ice flux is usually out of the Arctic Ocean, with a maximum of 152 x 10^3 km^2 in 1996. The years 1999 and 2004 are exceptions when ice flux is directed into the Arctic Ocean. The summer area flux is normally directed out of the Arctic Ocean, with a minimum of 30 x 10^3 km^2 in 1982. In the years 1979, 1981 and 1996 there was an ice flux in the opposite direction. The total area flux through the strait between Franz Josef Land and Novaya Zemlya (Transect 3) is positive for the entire period, with a minimum of 1.4 x 10^3 km^2 in 1980 and a maximum of 318 x 10^3 km^2 in 1997. The winter ice flux has a maximum of 292 x 10^3 km^2 and a minimum of 4 x 10^3 km^2. In summer there are significant variations in the flux direction from year to year.

Annual ice flux through the Transect 4 and Transect 5 during the period of observation is directed into the Arctic Ocean (negative), except for summer months in several years. Ice flux through Transect 4 varies from a minimum of -5.2 x 10^3 km^2 in 1998 (winter flux is -14 x 10^3 km^2) to a maximum of -229 x 10^3 km^2 in 2000 (winter flux is -230 x 10^3 km^2). Through Transect 5 the ice flux varies from a minimum of -123 x 10^3 km^2 in 1979 (winter flux is -113 x 10^3 km^2) to a maximum of -360 x 10^3 km^2 in 2002 (winter flux is -342 x 10^3 km^2). Winter months contribute significantly to the annual fluxes for both these transects, having maximum and minimum values the same years as the annual fluxes.

The ice area flux through Transect 6 has a high degree of inter-annual variability, changing its direction and values from year to year but on average is directed into the Arctic Ocean. The total area flux to the Arctic Ocean (negative) has a minimum in 1984 (-15 x 10^3 $km^{2)}$ and a maximum in 1996 (-179 x 10^3 km^2). Positive fluxes vary from 12 x 10^3 km^2 in 1988 to 88 x 10^3 km^2 in 1983. The winter area flux has the same behaviour as the total, on average it has the same direction, and ranges from about 73 x 10^3 km^2 in 1999 and 2005 (from the Arctic Ocean) to -188 x 10^3 km^2 in 1996 (to the Arctic Ocean). Summer area flux varies its direction from year to year. The ice area flux in Transect 7 also has a complicated inter-annual variability. The total area flux directed out of the Arctic Ocean has a maximum in 1994 (127 x 10^3 $km^{2)}$, and the flux directed into the Arctic Ocean has a maximum of -41 x 10^3 km^2 in 1996. The winter area flux is directed out of the Arctic Ocean and ranges from a minimum of 700 km^2 in 1981 to a maximum of 142 x 10^3 km^2 in 1994. Only in 1996 does the winter ice flux have a negative direction, out of the Arctic Ocean (-46 x 10^3 km^2 y^{-1}). Summer area flux during this period is mostly directed into the Arctic Ocean (except in 1980, 1996 and 2002), and ranges from -1 x 10^3 km^2 in 1981 and 1987 to -51 x 10^3 km^2 in 1988.

From the analysis of inter-annual variability we conclude that to all intents and purposes in Transects 1 to 5 the ice fluxes do not change direction during the period of the calculation (1979-2005). In the other two straits (Transect 6 and Transect 7) the ice flux changes direction in some years.

The sea ice area flux through Transects 1, 3, 4, 5 and 6 increases significantly during the 27-year observational period (see Figure 18). In particular, ice flux increased by 33% through Transect 1, 75% through Transect 3, 67% through Transect 4, and 51% through Transect 5. The increase of ice flux through Transects 2 and 7 is much less; 26% for Transect 7, and 5% for Transect 2. So we see that the marginal Siberian seas (Kara, Laptev and East Siberian seas) at present export much more ice to the Arctic Basin than in the previous decades. Together with Arctic warming, this can be the cause of the dramatic decrease in summer ice extent in these seas in recent years.

Table 1 compares our recent estimates of mean ice fluxes through Transects 1 to 7 with previous estimates. Previous estimations of the ice fluxes through these transects have been made by several authors for different periods and seasons, and using different approaches. Thus the results of these estimations are varied, but the results of our study are nearest to other estimations also based on satellite data (Kwok and Rothrock, 1999; Kwok, Cunningham and Pang, 2004; Kwok, Maslowski and Laxon, 2005).

Table 1. Comparison of annual flux estimates through the transects

Transect	Period	Area flux (10^3 km^2/year)	Sources
Transect 1	1979-2005	662	Present study
		650	Frolov et al. (2005)
	1978-2002	866	Kwok et al. (2004)
	1966-2000	639	Pavlov et a. (2004)
	1978-1996	919	Kwok and Rothrock (1999)
	1988-1994	584	Korsnes et al. (2002)
	1990-1996	1100	Vinje et al. (1998)
	1993-1994	715	Martin (1996)
	1976-1985	830	Thomas and Rothrock (1993)
	1979-1984	840	Moritz (1988)
	1976-1984	1210	Vinje and Finnekasa (1986)
	1967-1977	750	Vinje (1982)
Transect 2	1979-2005	78	Present study
		-5	Frolov et al. (2005)
	1994-2003 (Oct-Mar)	20	Kwok et al. (2005)
	1966-2000	70	Pavlov et a. (2004)
	1988-1994	75	Korsnes et al. (2002)
	1978-1996(Oct-May)	47	Kwok (2000)
Transect 3	1979-2005	158	Present study
	1966-2000	104	Pavlov et al. (2004)
	1988-1994	86	Korsnes et al. (2002)
	1978-1996(Oct-May)	126	Kwok (2000)
Transect 4	1979-2005	-78	Present study
		-120	Frolov et al. (2005)
	1966-2000	-87	Pavlov et al. (2004)
	1978-1996(Oct.-May)	-65	Kwok (2000)
Transect 5	1979-2005	-228	Present study
		-360	Frolov et al. (2005)
	1966-2000	-180	Pavlov et al. (2004)
Transect 6	1979-2005	27	Present study
	1966-2000	65	Pavlov et al. (2004)
Transect 7	1979-2005	31	Present study
	1966-2000	7	Pavlov et al. (2004)

CONCLUSION

As a result of this study we conclude that variability of the SID velocities on the annual time scale is controlled by seasonal changes of the wind. The maximum amplitudes of the annual cycle are found in the Fram Strait (9-10 cm s^{-1}), the Beaufort Gyre (6-7 cm s^{-1}) and the northern part of the Barents Sea (5-6 cm s^{-1}).

Lower frequency variations of SID velocities (2.0-2.5 yrs and 5.0-6.0 yrs) are related to reorganization of the atmospheric circulation over the Arctic. By analysing the inter-annual variability of the SID velocities we find a clear positive linear trend. In the Baffin Bay, Fram Strait and Barents Sea regions SID velocities have dramatically increased (0.15-0.20 cm s^{-1} per year) during the last two decades. We suggest that the increase of the SID velocities in the Arctic Ocean is mostly related to the decrease of both SIC and ice thickness.

The character of the inter-annual variability of ice exchange between the marginal Arctic seas and Arctic Basin and in the Fram Strait in general is similar to the variability of SID velocities averaged for the whole Arctic Ocean. We have found an increase (50-70%) in the ice export from Siberian seas to the Arctic Basin.

ACKNOWLEDGMENTS

Support for this work was partly provided by the Project "Sea ice data for the Greenland and Barents seas" (financed by the StatoilHydro, Contract No. 540 31 31, between NPI, NTNU and StatoilHydro).

REFERENCES

Buynitskiy, V.Kh. (1951). Formation and drift of the ice cover in the Arctic Ocean. *The Drifting Expedition on the Icebreaker G. Sedov in 1937-1940.* Vol.4, (pp. 74-151). Moscow, USSR: Glavsevmorput, (In Russian).

Cavalieri, D., Parkinson, C., Gloersen, P., and Zwally, H. J. (1996), updated 2006. Sea ice concentrations from Nimbus-7 SMMR and DMSP SSM/I passive microwave data, [lists dates of temporal coverage used]. Boulder, Colorado USA: National Snow and Ice Data Center. Digital media.

Chernyak, S., Rice, C., and McConnell, L. (1996). Evidence of currently-used pesticides in air, ice, fog, seawater and surface microlayer in the Barents and Chukchi seas. *Marine Pollution Bulletin, 5(32),* 410-419.

Colony, R., and Thorndike, A.S. (1984). An estimate of the mean field of Arctic Sea ice motion. *Journal of Geophysical Research, 89(C6),* 10623-10629.

Comiso, J.C. (2006). Abrupt decline in the Arctic winter sea ice cover. *Geophysical Research Letters, 33,* L18504, doi: 10.1029/2006GL027341.

Comiso, J.C. (2002). A rapidly declining perennial sea ice cover in the Arctic. *Geophysical Research Letters, 29(20)*, 1956, doi: 10.1029/2002GL015650.

Comiso, J. (1990), updated 2005. DMSP SSM/I daily and monthly polar gridded sea ice concentrations, [list dates used]. Edited by Maslanik, J., Stroeve, J. Boulder, Colorado USA: National Snow and Ice Data Center. Digital media.

Comiso, J.C., Parkinson, C.L., Gersten, R., and Stock, L. (2008). Acceleration decline in the Arctic sea ice cover. *Geophysical Research Letters, 35*, L01703, doi: 10.1029/2007GL031972.

De Long, G. W. (1883). *The voyage of the Jeannette: the ship and ice journals of George W. De Long, Lieutenant-commander U.S.N., and commander of the Polar expedition of 1879-1881.* Eedited by his wife, Emma De Long. Boston: Houghton Mifflin.

Fowler, C. (2003), updated 2007. Polar Pathfinder Daily 25 km EASE-Grid Sea Ice Motion Vectors. Boulder, Colorado USA: National Snow and Ice Data Center. Digital media.

Francis, J.A., Hunter, E., Key, J.R., and Wang, X. (2005). Clues to variability in Arctic minimum sea ice extent. *Geophysical Research Letters, 32*, L21501, doi: 10.1029/2005GL024376.

Frolov, I.E., Gudkovich, Z.M., Radionov, V.F., Shirochkov, A.V., and Timokhov, L.A. (2005). *The Arctic Basin, Results from the Russian Drifting Stations.* Chichester, UK: Praxis Publishing Ltd.

Gascard, J.C. (2007). Tara Damocles Press Conference, 30th of October 2007, press release. (http://www.taraexpeditions.org)

Gerland, S., Aars, J., Bracegirdle, T., Carmack, E., Hop, H., Hovelsrud, G.K., Kovacs, K., Lydersen, C., Perovich, D.K., Richter-Menge, J., Rybråten, S., Strøm, H., and Turner, J. (2007). Ice in the Sea. Chapter 5 In *Global outlook for Ice and Snow*. United Nations Environment Programme, Nairobi, (pp. 63-96).

Gudkovich, Z.M. (1961). Relationship of ice drift in the Arctic Basin with ice conditions in the Arctic seas. *Proceedings of Oceanographic Commission of the USSR Academy of Sciences, 11*, 13-20, (In Russian).

Hilmer, M., and Jung, T. (2000). Evidence of a recent change in the link between the North Atlantic Oscillation and Arctic sea ice export. *Geophysical Research Letters, 27(7)*, 989-992.

Hop, H., and Pavlova, O. (2008). Distribution and biomass transport of ice amphipods in drifting sea ice around Svalbard. *Deep-Sea Research II* (Accepted).

Hop, H., Falk-Petersen, S., Svendsen, H., Kwasniewski, S., Pavlov, V., Pavlova, O., and Søreide, J.E. (2006). Physical and biological characteristics of the pelagic system across Fram Strait to Kongsfjorden. *Progress in Oceanography, 71*, 182-231.

Karklin, V.P. (1977). State and future of using the geliogeophysical factors for ice forecusts. *Problems of Arctic and Antarctic, 50*, 45-48, (In Russian).

Kolchak, A.V. (1909). Ice of the Kara and the Siberian seas. *Notes of Russian Imperators' Academy of Sciences, 26(1)*, (In Russian).

Korsnes, R., Pavlova, O., and Godtliebsen, F. (2002). Assessment of potential transport of pollutants into the Barents Sea via sea ice–an observational approach. *Marine Pollution Bulletin, 44(9)*, 861-869.

Kwok, R. (2000). Recent changes in Arctic Ocean sea ice motion associated with the North Atlantic Oscillation. *Geophysical Research Letters, 27*, 775-778.

Kwok, R., and Rothrock, D.A. (1999). Variability of Fram Strait ice flux and North Atlantic Oscillation. *Journal of Geophysical Research, 104(C3),* 5177-5189.

Kwok, R., Maslowski, W., and Laxon, S.W. (2005). On large outflows of Arctic sea ice into the Barents Sea. *Geophysical Research Letters, 32,* L22503, doi:10.1029/2005GL024485.

Kwok, R., Cunningham, G.F., and Pang, S.S. (2004). Fram Strait sea ice outflow. *Journal of Geophysical Research, 109 (C01),* C01009, doi:10.1029/2003JC001785.

Lepparanta, M. ((2005). *The Drift of Sea Ice.* Chichester, UK: Praxis Publishing Ltd.

Lohmann, G., and Gerdes, R. (1998). Sea ice effects on the Sensitivity of the Thermohaline Circulation in simplified atmosphere-ocean-sea ice models. *Journal of Climate, 11,* 2789-2803.

Martin, T. (1996). Sea ice drift in the East Greenland Current. *Proceedings of Fourth Symposium on Remote Sensing of the Polar Environments, European Space Agency Special Publication, ESA SP-391,* 101-391.

Martin, T., and Gerdes, R. (2007). Sea ice drift variability in Arctic Ocean Model Intercomparison Project models and observations. *Journal of Geophysical Research, 112,* C04S10, doi:10.1029/2006JC003617.

Moritz, R.E. (1988). The ice budget of the Greenland Sea. *Technical Report.* APL-UW TR 8812, Seattle.

Mauritzen, C., and Hakkinen, S. (1997). Influence of Sea Ice on the Thermohaline Circulation in the Arctic-North Atlantic Ocean. *Geophysical Research Letters, 24(24)*, 3257-3260.

Nansen, F. (1903). *The Norwegian North Pole Expedition, 1893-1896, Scientific Results.* New York: Greenwood Press (reissued in 1979).

Pavlov, V. (2006). Modelling of long-range transport of contaminants from potential sources in the Arctic Ocean by water and sea ice. In J.B. Orbaek, T. Tombre, R. Kallenborn, E. Hegseth, S. Falk-Petersen, and A.H. Hoel (Eds.), *Arctic-alpine Ecosystems and People in a Changing Environment* (329-350). Berlin: Springer Verlag.

Pavlov, V.K., and Pavlova, O. A. (2007). Increasing sea ice drift velocities in the Arctic Ocean, 1979-2005. *Geophysical Research Abstract, Vol 9*, EGU-2007 General Assembly, 15-20 April, 2007, Vienna.

Pavlov, V., and Stanovoy, V. (2001). The problem of transfer of radionuclide pollution by sea ice. *Marine Pollution Bulletin, 4,* 319-323.

Pavlov, V., Pavlova, O., and Korsnes, R. (2004). Sea ice fluxes and drift trajectories from potential pollution sources, computed with a statistical sea ice model of the Arctic Ocean. *Journal of Marine Systems, 48,* 133-157.

Pfirman, S., Kogler, J., and Rigor, I. (1997a). Potential for rapid transport of contaminants from the Kara Sea. *The Science of the Total Environment, 202*, 111-122.

Pfirman, S, Colony, R., Nurnberg, D., Eicken, H., and Rigor, I. (1997b). Reconstructing the origin and trajectory of drifting Arctic sea ice. *Journal of Geophysical Research, 102(C6),* 12575-12586.

Pfirman, S.L., Eicken, H., Bauch, D., and Weeks, W. (1995). The potential transport of pollutants by Arctic sea ice. *The Science of the Total Environment, 159*, 129-146.

Pfirman, S., Haxby, W.F., Colony, R., and Rigor, I. (2004). Variability in Arctic sea ice drift. *Geophysical Research Letters, 31*, L16402, doi:10.1029/2004GL020063.

Proshutinsky, A., and Johnson, M. (1997). Two circulation regimes of the wind-driven Arctic Ocean. *Journal of Geophysical Research, 102(C6),* 12493-12514.

Rigor, I.G., Wallace, J.M., and Colony, R.L. (2002). Response of Sea Ice to the Arctic Oscillation. *Journal of Climate, 15*, 2648-2663.

Rothrock, D.A., and Zhang, J. (2005). Arctic Ocean sea ice volume: What explains its recent depletion? *Journal of Geophysical Research, 110*, C01002, doi:10.1029/2004JC002282.

Rothrock, D.A., Yu, Y., and Maykut, G.A. (1999). Thinning of the Arctic sea-ice cover. *Geophysical Research Letters, 26*, 3469-3472.

Serreze, M.C., Holland, M.M., and Stroeve, J. (2007). Perspectives on the Arctic's shrinking sea-ice cover. *Science, 315*, 1533-1536.

Serreze, M.C., Maslanik, J.A., Scambos, T.A., Fetterer, F., Stroeve, J., Knowles, K., Fowler, C., Drobot, S., Barry, R.G., and Haran, T.M. (2003). A record minimum arctic sea ice extent and area in 2002. *Geophysical Research Letters, 30*, 1110, doi:10.1029/2002GL016406.

Stroeve, J.C., Serreze, M.C., Fetterer, F., Arbetter, T., Meier, W., Maslanik, J.A., and Knowles, K. (2005). Tracking the Arctic's shrinking ice cover: Another extreme September minimum in 2004. *Geophysical Research Letters, 32*, L04501, doi:10.1029/2004GL021810.

Thorndike, A.S., and Colony, R. (1982). Sea ice motion in response to geostrophic winds. *Journal of Geophysical Research, 87(C8),* 5845-5852.

Thomas, D.R., and Rothrock, D.A. (1993). The Arctic Ocean ice balance: A Kalman smoother estimate. *Journal of Geophysical Research, 98(C6),* 10053-10067.

Vinje, T. (1982). The drift pattern of sea ice in the Arctic with particular reference to the Atlantic approach. In L. Rey and B. Stonehouse. (Eds.), *The Arctic Ocean* (pp. 83-96). Indianapolis: Macmillan.

Vinje, T., Nordlund, N., and Kvambekk, A. (1998). Monitoring ice thickness in Fram Strait. *Journal of Geophysical Research, 103(C5):*10437-10449.

Vinje, T., and Finnekasa, O. (1986). The ice transport through Fram Strait. *Report, 186*, 1-39. Norsk Polarinstittut, Oslo, Norway.

Vorobyev, V.N., and Gudkovich, Z.M. (1976). Inter-annual variability in the ice drift and currents of the Arctic Basin. *Proceeding of the AARI, 319*, 23-32, (In Russian).

Walsh, J.E., Chapman, W.L., and Shy, T.L. (1996). Recent decrease of sea level pressure in the central Arctic. *Journal of Climate, 3,* 1462-1473.

Weeks, W. (1994). Possible roles of sea ice in the transport of hazardous material. *Arctic Research of the United States, 8*, 34-52.

Zhang, J.L., Rothrock, D., and Steele, M. (2000). Recent changes in Arctic sea ice: The interplay between ice dynamics and thermodynamics. *Journal of Climate, 13,* 3099-3114.

Zubov, N.N. (1943). *Arctic Ice*. Moscow: Glavsevmorput. Translated by US Navy Oceanographic Office, Springfield.

INDEX

A

B

C

D

E

F

G

M

N

O

P

T

U

V

W

Z